TRACTION ENGINE MUSEUM GUIDE

BARRIE C. WOODS

ACKNOWLEDGEMENTS

Production of this directory would not have been possible without the help and support of numerous people and organisations. In particular I would like to thank Derek Wheeler, MBE; Brian Sharp, Chairman of the Bedfordshire Steam Engine Preservation Society; Brian Johnson, Editor of the Traction Engine Register; John Cook of the NTET and the many museums and other organisations who have contributed. **Barrie C. Woods: January 2010.**

Front cover: *The East Anglian Transport Museum at Carlton Colville, near Lowestoft is host to this Armstrong Whitworth Roller Works No. 10R22, seen at the Grand Henham Rally on 20th September 2008.*

Published by:
Barrie C. Woods
January 2010

Copyright: Barrie C. Woods, 12th December 2009.
For further information on this matter please contact:
Barrie C. Woods, 124 Eastern Way, Letchworth Garden City.SG6 4PF.
01462-634835. E.mail: bcwoods124@ntlworld.com

Except for normal review purposes, no part of this book may be reproduced or transmitted in any form or by any means, electronic or mechanical, including photocopying, recording or by any information storage and retrieval system without the written permission of the Author.

Robey Tri-Tandem roller No. 45655 Reg: VL2773 is currently a resident at the Robey Trust works in Tavistock. Built in 1930, one of her last jobs was working on the M1 motorway. Named 'Herts Wanderer' she is seen here at the Pirton Traction Engine Rally in 1975.

Foreword

In over 40 years of being an exhibitor at steam rallies all over England, I thought I had seen most of the engines on the regular circuit. It came as a pleasant surprise to find that, thanks to Barrie's extensive research, many more engines are to be found in museums all over these islands.

As the owner of a Shand Mason steam fire engine, my suspicions were confirmed that there are more of these handsome vehicles static in museums and fire stations than the public ever sees in action on a rally field. This little book has whetted my appetite to explore further and seek out elusive machines by other makers too.

It is a volume that fills a niche in the compendium of transport literature and Barrie must be congratulated for recognising this long-felt need and addressing the problem so admirably.

Derek Wheeler MBE

Derek Wheeler MBE, resplendent in full period uniform with his beautifully restored and fully operational Shand Mason 'St George', at the Great Dorset Steam Fair in 2009.

TRACTION ENGINE MUSEUM GUIDE

BY BARRIE C. WOODS

(2010)

Listing of all traction engines, steam rollers, steam fire engines and steam wagons housed in museums and other locations throughout the UK and Ireland.

Contents

Introduction	7
Index to museums and other locations	10
Main Museum directory	14
Engine cross-reference section	91
Museums by County	102

The Traction Engine Register is a guide to the age, identity, technical details and location of all the known surviving traction engines, road locomotives, steam wagons, road rollers and fire engines that survive in the United Kingdom and Ireland.

Published every four years in an updated edition - the latest in 2008 - it lists approximately 3000 engines. Included is a reference guide that identifies the engine maker and maker's number from the road registration - a particularly useful feature for identifying engines from photographs.

Also included is a reference to engines that have been exported in recent years whilst in preservation and a selection of colour and monochrome pictures illustrating the various principle types of road engines. A listing of the main road steam societies is provided. Often referred to as the 'enthusiasts bible' it is an invaluable guide to anyone interested in road steam

The Traction Engine Register is available from the publishers: Southern Counties Historic Vehicles Preservation Trust, 2c Hevers Avenue, Horley, Surrey RH6 8DB, price £8.50 including postage, or from club and museum sales shops.

TRACTION ENGINE MUSEUM GUIDE

Introduction

The well established 'Traction Engine Register' continues to be an excellent source of information with regard to the traction engines extant in the United Kingdom. The comprehensive, concise and detailed listing of almost all engines around the country has proven this publication to become the 'Bible' for any enthusiast.

However, the book does not effectively cover those engines in the many museums scattered across the land. There are some brief references to them but it has not been that publication's mandate to detail this aspect of the movement. Thus there is a specific market interest that up to now has not been accommodated for that information, which is laid out herein.

The object of this booklet is to inform the reader of all the museums and other locations in the country that either primarily or secondarily house traction engines, steam wagons and associated steam road-going vehicles. Museums such as Thursford, Strumpshaw and Bressingham are obvious candidates for inclusion. Other locations are not so well known such as Farmer Parr's in Fleetwood, Christie's Garden Centre at Fochabers or others that may appear to house other types of exhibits but which include one of two engines like the Chain Bridge Honey farm near Berwick or Dunrobin Castle where there is a Shand Mason fire engine tucked away in the café!

Museums by their very title vary enormously in the way they display exhibits. Some will have rows of polished and lovingly cared for engines; others may well have a rusting little engine tucked away in a corner. Yet more have engines listed which in fact at the time of writing are not on display and may be in the workshop or indeed at another 'store' location nearby. Birmingham Science Museum is an example of the latter. Where some engines are stored or not on display, it is only correct and courteous to enquire as to whether access can be gained to view or photograph them. Refusal of this request may be for a number of reasons; health & safety, staff shortage, lack of keys and so on. These refusals should always be accepted courteously as on another day your request may well be accepted.

A number of Fire Stations are included in this list; by the nature of their activities it is essential to ensure you have prior permission to view their engines. These premises are not museums but active work places with all the health and safety aspects that go along with them. From my own experiences a very warm welcome is always the case at such locations.

Following considerable research I have established and included herein a total of 458 engines of all types, including a couple of diesel and paraffin engines at no less than 146 locations. These are spread far and wide although some towns or cities may have up to three locations where engines may be found.

Whatever your particular interest in this vast hobby I hope this booklet will help you locate and enjoy the exhibits and indeed the many other non-steam items on display in these museums. This is an ever-changing situation and if you see something that is incorrect or can add anything to the next issue please do get in touch with me. Photographs to back up your sightings are always welcome as well. Above all, enjoy yourselves.
Barrie C. Woods.

How to use this book
Full details of museums follow, including directions, postal details, e.mail address and web-site where possible. Post-codes are included for Sat-Nav users. A cross-reference of engines from page 91 by builder will allow the reader to check where any particular engine is located.

Important notices:

Note 1:
There are of course several other notable collections of engines such as Barrack's at Aberdeen, Crawford's at Frithville and Saunders at Stotfold. These collections are not normally open to the public and are constantly changing, therefore not included in this analysis. Whilst some of the residents at Preston Services, Canterbury may be considered 'permanent' details of this collection are not included as the operation is primarily a commercial business and thus engines frequently move on. Conversely Klondyke Mill, which is not a museum as such, is the base for the North Staffordshire Traction Engine club and access may be possible with permission. This factor is relevant for other locations such as Tinkers Park, and Sellinge; both sites have resident engines but they may only be available to study on special events such as an annual traction engine rally.

There are others such as Chatham Dockyard and Bursledon Brickworks which again have restricted access. Please ensure you abide by any criteria laid down by any organisation.

Note 2:
This list is collated on the basis of information available from the organisations at the time of compiling this edition and from my own observations. It in no way constitutes or is even meant to constitute a full list of museum engines. No responsibility is accepted by the author of this list for any inaccuracies herein. An entry herein does not give anyone permission to enter any of the establishments listed, the details about engines, times of opening etc are only for information purposes. Publication or distribution of part or all of the information in this list in any form either by hard copy or electronic mail is forbidden and regulated by copyright laws to the author.

Note 3:
Up to date or additional information which can be included in a future edition is always welcome, but should only be submitted following personal observations, backed up with a photograph wherever possible, not via hearsay.

Note 4:
Some engines located at museums are privately owned and thus may well be moved or re-located at any time. Even with museum owned vehicles these can and are often moved from one site to another for various reasons, or may be out on a rally at the time you visit, or even be placed in store and not available to see. Milestones Museum in Basingstoke is a case in point where engines are exchanged with Chilcomb House, Winchester from time to time.

Note 5:
As can be imagined the information contained within this booklet is ever-changing, thus I intend to update as necessary in order to keep up with the latest movements of engines.

Note 6:
All photographs are from the author's collection unless otherwise stated.

Index to museums and other locations

Museum/location	Page No.
Aberfeldy, Victoria Park, Perth & Kinross	14
Alfold, Grampian Transport Museum, Aberdeen.	14
Amberley Chalk Pits Museum, West Sussex	15
Angmering, Haskins Garden Centre, West Sussex	15
Arreton Barns Craft Village & Museum, Isle of Wight	16
Aylesbury, Quainton Road, Buckinghamshire	16
Banbury, Bygones Museum, Oxfordshire	17
Bangor, Penrhyn Castle Museum, Gwynedd	17
Barleylands Farm Museum, Essex	17
Barnsley, Wortley Top Forge Industrial Museum	18
Basingstoke, Milestones Museum, Hampshire	18
Beamish Open Air Museum, Durham	19
Beaulieu, National Motor Museum, Hampshire	20
Belfast, Ulster Folk & Transport Museum, Co. Antrim	20
Berwick-on-Tweed, Chain Bridge Honey Farm, Northumberland	21
Bicton Gardens, Devon	21
Biggleswade Fire Station, Bedfordshire	21
Birmingham Science Museum, Dollman Street, West Midlands	22
Birmingham Science Museum, Thinktank, West Midlands	23
Birmingham, West Midlands Fire & Rescue Service	23
Blackgang Chine, Nr. Ventnor Isle of Wight	23
Blaenafon, Big Pit National Coal Museum	24
Bradford Industrial Museum	24
Brading, Isle of Wight, 'The Experience'	25
Brentwood, Old MacDonald's Farm, Essex	25
Bridport, Highlands End Holiday Park, Dorset	26
Brighton, British Engineerium, Brighton & Hove	26
Bristol Industrial Museum, Bristol	26
Brownhills, Chasewater Railway, West Midlands	31
Burton-on-Trent, Coors Museum, Staffordshire	31
Burton-on-Trent, Staffordshire Fire & Rescue Museum	31
Bury, East Lancashire Railway Museum, Lancashire	32
Caister Castle Museum, Norfolk	32
Calbourne Mill, Isle of Wight	32
Cambridge Museum of Technology	33
Canterbury, Preston Services, Kent	33
Cardiff, Welsh Industrial & Maritime Museum, St. Fagans, Cardiff	33
Carrickfergus Museum & Civic Centre, Co. Antrim	33
Chatham Steam Centre, The Historic Dockyard, Medway	34
Chester, Cheshire Fire & Rescue Service, Cheshire	35
Chesterfield, Barrow Hill Roundhouse Railway Centre, Derbyshire	35

Chinnor & Princes Risborough Railway, Oxfordshire 35
Chippenham, Lackham Agricultural & Rural Life Trust, Wiltshire 36
Coalville, Snibston Discovery Park, Leicestershire 36
Coatbridge, Summerlee, Scottish Museum of Industrial Life, N.Lanarks .. 37
Denby Dale, Sisset, Nortonthorpe Mills, West Yorkshire 37
Dereham, Gressenhall Farm & Workhouse Museum, Norfolk 37
Didcot Railway Centre, Oxfordshire 38
Diss, Bressingham Steam Museum, Norfolk 38
Doncaster, Cusworth Hall and Park, South Yorkshire 39
Dorchester, Recreational Park, Dorset 39
Dover Transport Museum, Kent 39
Draycott-in-the-Clay, Klondyke Mill, Staffordshire 40
Dublin, Straffan Steam Museum, Co. Dublin 40
Dublin, The National Transport Museum, Howth Castle, Co. Dublin 41
Dunfermline, Lathalmond Bus Museum, Fife 41
Dunrobin Castle, Golspie, Highland 41
Edinburgh, Lothian and Borders Fire and Rescue Service 42
Edinburgh, National Museums Collections Centre, Granton, Midlothian .. 42
Edinburgh, Royal Museum of Scotland 43
Ely Fire Station, Cambridgeshire 43
Enfield, Whitewebbs Museum of Transport, London 43
Exmouth, Sandy Bay, World of Country Life, Devon 44
Fakenham, Thursford Collection, Norfolk 44
Fleetwood, Farmer Parrs Animal World, Lancashire 46
Fochabers, Christies Garden Centre, Morayshire 46
Fordingbridge, Breamore House, Hampshire 47
Forres, Dallas Dhu Historic Distillery 47
Glasgow Museum of Transport, Glasgow 47
Gloucester City Museum, Gloucestershire 48
Greenock, Strathclyde, Fire Preservation Group,Strathclyde 48
Harlington, Poplars Garden Centre, Bedfordshire 49
Helston, Flambards, The Experience, Cornwall 49
Hollycombe House, Steam in the Country, West Sussex 49
Ipswich, Suffolk Fire Service, Suffolk 50
Ironbridge Gorge Museum, Blists Hill, Shropshire 51
Jersey, Pallot Steam, Motor & General Museum, Jersey 51
Kendal, Levens Hall, Cumbrla 51
Kenilworth, Stoneleigh Abbey, Warwickshire 52
Kew Bridge Steam Museum, London 52
Kidderminster, Hodgehill Nurseries, Worcestershire 53
Knutsford, Tatton Hall, Cheshire 53
Launceston, Dingles Fairground Heritage Centre, Cornwall 53
Launceston Steam Railway, Cornwall 54
Leeds, Armley Mills Industrial Museum, West Yorkshire 54

Leicester Museum of Technology, Leicestershire..................................59
Leiston, Long Shop Museum, Suffolk 59
Leyland, British Commercial Vehicle Museum, Lancashire 60
Lincoln, Museum of Lincolnshire Life, Lincolnshire................................60
Liverpool, Museum of Liverpool 61
Lowestoft, East Anglian Transport Museum. Suffolk 61
Luton, Stockwood Museum, Bedfordshire 62
Maidstone, Kent Fire Service, Kent..62
Maidstone Museum & Bentlif Art Gallery, Kent..62
Midleton, The Old Distillery, Co. Cork, Ireland 63
Milton Keynes Rural Life Museum, Buckinghamshire..............................63
Moreton-in-the-Marsh, Fire Service College, Oxfordshire 64
Newby Bridge, Lakeside & Haverthwaite Railway, Cumbria....................64
Newcastle Discovery Museum, Tyne & Wear..65
Northallerton, Prestons of Potto, North Yorkshire....................................65
Norwich, Bridewell Museum, Norfolk 66
Nottingham, Woollaton Park, Nottinghamshire 66
Oxford City Fire & Rescue Service, Oxfordshire......................................66
Peastonbank, Glenkinchie Distillery, East Lothian...................................67
Poole Museum, Poole 67
Portsmouth City Museum, Portsmouth 68
Preston, Lancashire Fire & Rescue Service, Lancashire...........................68
Ramsey Rural Museum, Cambridgeshire 68
Reading, Englefield Estate, Berkshire..68
Reading, Museum of English Rural Life, Berkshire 69
Rochdale, Greater Manchester Fire Service Museum...............................69
St Austell, on private land alongside the A390, Cornwall 70
Saltash, St. Dominick, Cotehele House, Cornwall 70
Scarborough, The Scarborough Fair Collection. North Yorkshire............70
Scunthorpe, Normanby Hall Country Park, Lincolnshire.........................71
Sheffield Fire & Police Museum, South Yorkshire 71
Sheffield, Kelham Island Museum, South Yorkshire................................71
Sheffield, Renishaw Hall, South Yorkshire..72
Sittingbourne & Kemsley Light Railway, Kemsley Station, Kent.............72
Sittingbourne, Bredgar & Wormshill Railway 73
Skegness, Church Farm Museum, Lincolnshire 73
Southampton, Bursledon Brickworks, Southampton................................74
Southwark, London Fire Brigade Museum, London.................................74
South Kensington, Science Museum, London 75
Stoke, Potteries Museum & Art Gallery, Stoke 75
Stowmarket, Museum of East Anglian Life, Suffolk 75
Stradbally Steam Museum, Laois, Eire 76
Strumpshaw Steam Museum, Norfolk..76
Swaffham, Cockley Cley, Iceni Village..77

Swansea, South Wales Fire & Rescue Service, Swansea 77
Swansea Waterfront Museum, Swansea 78
Swansea, Welsh Museum of Fire, Swansea 78
Tamworth, Statfold Barn Railway, Staffordshire 78
Tavistock, The Robey Trust, Devon 79
Thetford, Charles Burrell Museum, Norfolk 80
Thirsk, Ampleforth Farming Flashback 80
Watford Fire Station & Museum, Hertfordshire 81
Weedon, National Fire Service Museum, Northamptonshire 81
Wells-next-the-Sea, Holkham Hall Bygones Museum, Norfolk 82
Westonzoyland Pumping Station, Somerset 82
Weybourne, North Norfolk Railway, Norfolk 87
Winchester, Chilcomb House, Hampshire 87
Wroughton, Science Museum, Swindon 88
York Castle Museum, York 89
York, National Railway Museum, York 89

Possibly the most famous of all traction engines, 'The Iron Maiden'. Fowler No. 15657 of 1920. Now resident at the Scarborough Fair Collection and seen here at Battersea Park on 14th May 1972.

Museums and other sites with steam exhibits

This directory lists alphabetically to the nearest significant town or village, all known museums and other locations such as fire stations, parks and garden centres for instance, which include steam exhibits within their complex or as part of the main collection. To assist the visitor I have entered the full address, and post-code (to aid those with sat-nav). Directions are given from the nearest main road. Tel. No. E.mail and Web-sites are added where possible or relevant. In addition I have included opening times and admission details as well as any other relevant information as at the time of writing.

Take note; where specified, there are locations where the vehicle is listed but it is not on public display or there are restrictions as to viewing. These are included as situations do change and with the intent to make this as comprehensive a list as possible I deemed it important to at least know where the particular engine is located; who knows next year it may be on display! Please ensure you abide by any of these details. Information cited here was correct at the time of printing but the author takes no responsibility for any changes to opening times or admission charges to any museum, or indeed the availability of its exhibits.

The format of the information in the boxes is as follows:

Manufacturer.	Works No.	Registration.	Year built.	Name.	Type of engine

Aberfeldy, Victoria Park
This engine is located in the public park in the town.
Directions: From Perth take the A9 north for 12 miles to Little Dunkeld, after another 8 miles turn left on the A822 to Cablea and turn right on the A826, Aberfledy is 9 miles further on this road.

Fowler	16402	SW 1854	1925	Tillie	RR

Alfold, Grampian Transport Museum
Main Street, Alfold
Aberdeenshire
AB33 8AE
Directions: From Aberdeen take the A944 west for 25 miles to Alfold. The museum is by the public car park in Montgarrie Road, *(off the main street)* Alfold.
Tel: 019755-62292
www.gtm.org.uk
E.mail: info@gtm.org.uk

Opening times:
1st Apr - 31st Oct: 10.00 - 5.00pm
Other times: 10 - 16.00pm
Closed Christmas and New Year
Admission
£6.00 with Concessions

Aveling & Porter	14121	FG 7099	1931	Dragon	RR Class AE
Lawson	-	SA 16	1895	The Craigievar Express.	Steam Tricycle
Marshall	89200	-	1942	Birkhall	Portable
Sentinel	753	V 3057	1914	-	Wagon
Shand Mason	-	-	1894	GNSR Fire Brigade	FE

Amberley Working Museum
Amberley
Nr. Arundel
W. Sussex
BN18 9LT
Directions: The museum is a vast chalk pit adjacent to Amberley Railway station on the B2139 approximately 5 miles north of Arundel.
Tel: 01798-831370.
www.amberleymuseum.co.uk
E.mail: office@amberleymuseum.co.uk
Opening times;
14th Feb - 1st Nov: Tues - Sun: 10.00 - 17.00.
Admission:
£9.50 with Concessions.

Aveling Barford	AG 758	FRM 973	1946	Gisela	RR
Aveling & Porter	14090	NG 564	1931	Polly	RR
Marshall	68872	OT 3092	1915	-	TE
Marshall	79669	PX 2690	1925	Joan	RR
Marshall	80608	FG 1191	1926	Omon	RR
Shand Mason	-	-	1896	Warnham Court	FE *
Shand Mason	-	-	1895	Arundel Castle	FE **
Tasker	1396	AA 2254	1908	-	T
Wallis & Steevens	8104	GOR 248	1936	-	RR

* Not due to arrive until 2010/11
** Ex Worthing Fire Station

Angmering, Haskins Garden Centre
Roundstone by-Pass
Angmering
Littlehampton
W.Sussex
BN16 4BD

Directions: Located on the east-bound carriageway of the A259 Roundstone By-pass to the north of East Preston between Littlehampton and Worthing.
Tel: 01903-777666.
www.haskins.co.uk
E.mail: roundstoneenquiries@haskins.co.uk
Opening times: Daily but vary - see web-site. Closed Xmas day and Easter day.
Admission: Free.

Note: The engine can be seen from the by-pass.

Marshall	-	-	-	-	Portable

Arreton Barns Craft Village & Museum
Arreton
Main Road
Isle of Wight
PO30 3AA
Directions: Situated off the main Newport to Sandown Road the A3056. From Newport, proceed through Blackwater, about 2 miles on the road takes a sharp left, after another ½ mile a sharp right, the site is on the left about ¼ further on.
Tel: 01983-533079.
www.iowight.com/shipwrecks/themuseum.htm
E.mail: info@arretonbarns.co.uk
Opening times: Easter to 31st Oct: 10.30 - 16.30.
Admission: Free.

Marshall	69872	-	1916	Marion	Portable

Aylesbury, Quainton Road
Buckinghamshire Railway Centre
Quainton Road Station
Quainton
Nr Aylesbury
Bucks
HP22 4BY
Directions: This steam centre is 6 miles north-west of Aylesbury, head west on the A41 and turn right at Waddesden. Signs will lead you to the centre about 2 miles further on.
Tel:01296-655720.
www.bucksrailcentre.org
E.mail: office@bucksrailcentre.org
Opening times: April - Oct: Wed - Sun: 10.30 - 16.30 + Bank holidays.
Admission: See Web site.

| Aveling & Porter 803 | - | 1895? | - | Rail loco |

Note: On loan from the London Transport Museum

Banbury, Bygones Museum
Butlin Farm
Claydon
Nr Banbury
Oxon
OX17 1EP
Directions: Claydon is about 5 miles north of Banbury. From the A423 Southam Road turn east at sign-post 'Claydon' continue into village on Mollington Road, at 'T' Jct with Main Street turn right, the museum is on the left about 50 mtrs further on.
Tel: 01295-690258.
www.aboutbritain.com/BygonesMuseum.htm
E.mail: See Web-site
Opening times: April - Oct: Wed-Sun: 10.30 - 16.30.
Admission: £2.50 with concessions.

Note: The engine is owned by the Banbury Steam Engine Society

| Merryweather Fire Pump | - | - | - | Valiant type |

Bangor, Penrhyn Castle Museum
Bangor
Gwynedd
LL57 4HN
Directions: Located just west of Bangor off the A5, follow brown tourist signs.
Tel: 01248-353084.
www.nationaltrust.org.uk/hbcache/property140.htm
E.mail: penrhyncastle@nationaltrust.org.uk
Opening times: 22nd Mar - 31st Oct: Mon, Wed - Sun: 12.00 - 16.30.
Admission: See web-site.

| Merryweather | - | - | 1894 Penrhyn Castle | FE |

Barleylands Discovery Centre
Barleylands Farm,
Barleylands Road
Billericay, Essex
CM11 2UD
Directions: From the M25/A127 exit at Jct 29 on to A127 (Southend direction), come off at A176 for Basildon/Billericay, at roundabout take third exit on to South Wash Road follow brown tourist signs to 'Farm Museum'. At 2nd

roundabout turn left into Barleylands Road the Centre is along here on the right hand side.
Tel: 01268-290228.
www.barleylands.co.uk
E.mail: info@essexcountryshow.co.uk
Opening times: Mar - Oct: 10.00 - 17.00. Nov - Feb: 10.00 - 16.00.
Closed Christmas and New Year.
Admission: Farm £6. Craft village; free.

Fowler	14728	DO 1918	1918	Giant Tiger	PLG
Fowler	14729	DO 1919	1918	Giant Panther	PLG
RSJ	42019	-	1931	The Wingham	Portable

Barnsley, Wortley Top Forge Industrial Museum
Forge Lane
Thurgoland
South Yorks
S35 7DN
Directions: This museum is located south-east of Barnsley. From the M1 turn west at Junction 36 on to the A61 at the roundabout with the A616 turn right, then at cross roads with the A629 turn right, (signposted Huddersfield and Wortley) after ½ mile pass under a low bridge and turn left into Finkle Street Lane. At the next 'T' junction turn right, Wortley Forge is ½ mile along this road.
OS Ref; SK 294998.
Tel: 0114-288-7576.
www.topforge.co.uk
Opening times: Sundays only Feb - Nov: 11.00 - 17.00.
Admission: £3.00 + Concessions.

| F. Clarke & Co | UP4 | - | 1880? | | Portable |
| Shand Mason | - | - | ? | - | Steam Pump |

Basingstoke, Milestones Museum
Leisure Park
Churchill Way West
Basingstoke, Hants
RG22 6PG
Directions: from M3 take Jct 6 and head north on A30 Ringway East, turn left on A3010 Churchill Way West go on to roundabout, continue straight over this to next large roundabout and go all the way round heading back from whence you came. The entrance to museum is on the left a short way along.
Tel: 0845-603-5635.
www.milestones.museum.com
E.mail: milestones.museum@hants.gov.uk
Opening times: Tues - Fri & Bank Hols 10.00 - 17.00. Sat-Sun: 11.00 - 17.00.

Closed Christmas and New Year.
Admission: Farm £7.50 with + Concessions.

Merryweather	1292	-	-	-	FE
Tasker	111	-	1872	-	Semi portable
Tasker	1228	-	1898	-	Portable
Tasker	1352	HO 5600	1905	-	TE
Tasker	1599	AA 5296	1914	-	Tr
Tasker	1643	AA5506	1915	-	Sectioned boiler only
Tasker	1726	SR 1294	1917	Blossom	Chain Tr
Tasker	1906	BD 7994	1923	-	RR Class B2
Tasker	1915	YB 183	1924	-	Wagon
Tasker	1933	OT 8201	1928	-	RR Class C
Thornycroft	115	EL 3908	1902	-	Wagon
Wallis & Steevens	7279	AA 2470	1912	-	Wagon
Wallis & Steevens	7867	OT 3078	1926	Old Lytham II	RR
Wallis & Steevens	7940	OT 8512	1928	Susie	RR

Beamish Open Air Museum
Beamish Museum
Beamish
Co. Durham
DH9 0RG
Directions: 9 miles northwest of Durham, Beamish is sign-posted off the A1M. Come off at Jct 63 (Chester-le-Street) on to A693, head for Stanley for 4 miles follow brown tourist signs.
Tel: 0191-370-4000.
www.beamish.org.uk
E.mail: museum@beamish.org.uk
Opening times:
Winter: 3rd Jan - 3rd Apr; Tues, Thurs & Weekends 10.00 - 16.00.
Summer: 4th Apr - 1st Nov, 7 days a week 10.00 - 17.00.
Admission:
Winter: £6.00.
Summer: £16.00 with concessions.

Aveling Barford	AH 363	CTL 225	1947	Linda	RR
Clayton & Shutt	13818	-	1874	-	Portable
Fowler	15490	CA 6328	1920	Fiddler	RR
Fowler	18877	WX 6358	1931	Rambler	RR
Mann	1747	MUP 662	1928	-	Tr
Richard Hornsby	2598	-	1875	-	Portable
Robey	51407	-	1947	-	Exploded boiler
Ruston Proctor	47319	-	1913	-	Portable
Savage	627	-	1895	May Queen	CE
Savage	713	-	1898	-	OE
Shand Mason	-	-	1907	City of York	Fire Engine
Shand Mason	-	-	1890	The Nelson	Bishop Auckland. FE

Beaulieu Motor Museum

Beaulieu
Brockenhurst
Hants
SO42 7ZN
Directions: From the M27 take Jct 2 and follow the brown tourist signs to the museum.
Tel: 01592-612345.
www.beaulieu.co.uk
E.mail: info@beaulieu.co.uk
Opening times: 23rd May - 30th Sept: 10.00 - 18.00.
1st Oct- 22nd May: 10.00 - 17.00.
Admission: £15.50 with concessions. Closed Christmas day.

Grenville	-	-	1875	-	Steam carriage
Merryweather	-	-	1915?	-	Valiant type*
* Note: Mounted on the rear of a 1907 Gobron Brillie Truck					

Belfast, Cultra, Ulster Folk & Transport Museum

153 Bangor Road
Holywood
BT18 0EU
Directions: Situated 7 miles east of Belfast near Holywood on the main A2 road to Bangor.
Tel: 028-9042-8428.
www.uftm.org.uk
E.mail: info@nmni.com
Opening times: Refurbishment is currently in progress at this museum. See web-site for re-opening date.
Admission: £6.50 with concessions.

Foster	14438	NR 7262	1921	No.105	(in store)
Fowler	15662	AZ 2450	1929	-	Class R3 (in store)
Fowler	19357	CZ 2260	1933	-	RR
Fowler	19590	JI 5869	1934	-	RR (in store)
Merryweather	-	-	-	The York St. Flax Spinning Co.	FE
W. Rose	-	-	1892	Belfast FB No.6	Fire Engine

Note: Of the three engines here only the W. Rose fire engine and Fowler 19357 are on display, the others are stored. Also note that part of the museum is being refurbished but is due to re-open in at the beginning of 2010. For more information on this: www.ulstermuseum.org.uk

Berwick-on-Tweed, Chain Bridge Honey Farm, Horncliffe
Horncliffe
Berwick-on-Tweed
TD15 2XT
Directions: The Farm is to the west of Berwick-on-Tweed. From the Berwick by-pass turn west on to the A698 for Coldstream, Turn right at Longridge Towers, head for Loanhead. Turn left at 'T' Jct on to Horncliffe Road for about ¾ mile, then turn right and the farm is 500 yards on the right.
Tel: 01289-386362.
www.chainbridgehoney.co.uk
E.mail: info@chainbridgehoney.co.uk
Opening Times: Nov - Mar: Mon - Fri: 09.00 - 17.00
Apr - Oct: 10.00 - 17.00 7 days a week.
Admission: Free.

Note: This is a commercial honey producing farm and not a museum; however once you have entered the premises the engine is easily accessible. (There are 2 ex London Transport Routemaster buses there also).

| Marshall | 53755 | SD 5453 | 1910 | - | RR Conv. |

Bicton Park Botanical Gardens
East Budleigh
Budleigh Salterton
Devon. EX9 7BJ
Directions: Come off M5 at Jct 30 take A3052 to Newton Poppleford then turn right on to B3178, 1 mile after Colaton Raleigh, Bicton Gardens is on the right just after the entrance to the college.
Tel: 01395-568465.
www.bictongardens.co.uk
E.mail: info@bictongardens.co.uk
Opening times:
Summer: 10.00 - 18.00
Winter: 10.00 - 17.00: Closed Christmas & New Year's Day.
Admission:
£6.95 with concessions.

Babcock & Wilcox	95/4011	YB 7978	1926	-	RR
Marshall	23368	TA 2758	1894	-	Diesel conversion
Marshall	86753	-	1932	-	Portable

Biggleswade Fire Station
Chestnut Avenue
Biggleswade, Beds. SG18 0LL

Directions:
From the A1 heading north come off at the roundabout at the south end of the town on to the A6001. After 100mtrs go straight over a roundabout and after 1 mile the road bears left and the B1040 goes off to the right, stay on the A6001 road until the next junction with the B1040, turn right on to it and immediately left into Chestnut Avenue. The Fire station is on the right after a few mtrs. If heading south, exit the A1 at the Sainsbury roundabout turning left into the town. Follow this for about ¾ mile, there is a sharp left followed by humps on the road, cross a mini roundabout and the railway bridge, after 300 mtrs the main road bears right. Carry straight on to the B1040, immediately turn left into Chestnut Avenue the fire station is on the right.
Tel:01234-351081 or 077424-33897 (ask for Eddie).
www.bedsfire.com
E.mail: contact@bedsfire.com
Opening times: There are usually steam enthusiasts present at the fire station on Monday evenings after 7.30pm, visitors are welcome to call in.
Admission: Free, but donations are welcome.

Shand Mason	2015	-	1908	Thorney	FE

Birmingham Science Museum, Dollman Street
Museums Collection Centre
25 Dollman Street
Nechells
B7 4RQ
Directions: Located near Duddeston Street railway station, or by bus number 26. By car it is south-east of the junction of the A47 and A4540. Going north on the A4540 turn right into Vauxhall Road, the B4132, take the fifth right into Erskine Street, go over railway bridge and turn left into Dollman Street, the museum is on the right. From the north, on the A4540 go straight over the Ashted Circus (roundabout with the A47), turn left into Vauxhall Road and proceed as above. From the east on A47 go along Saltley Road to roundabout with B4132, Melvina Road, (the railway is on your left), at 'T' Jct turn left into Duddeston Mill Road, go over railway then right into Dollman Street, the museum is on your left.
Tel: 0121-303-0910. For guided tours: 0121-303-2836.
www.birminghamroundabout.co.uk
E.mail: See web-site
Opening times: See below.
Admission: Check on arrival.

Note: *The Dollman Street store is only open for two days in the year, check web-site for details. Guided tours for specials parties are available at other times but it is essential to ring to make arrangements as indicated above.*

Aveling & Porter	2992	AB 9331	1892	-	Class R10
Burrell	4084	FK 3564	1927	Busy Bee	TE
Shand Mason	-	-	1898	-	Fire Engine
Shand Mason	-	-	1906	BSA II	Fire Pump
Shand Mason	-	-	-	-	Incomplete fire pump

Birmingham, Thinktank Science Museum
Millennium Point
Curzon Street
Birmingham
B4 7XG
Directions: The museum is located adjacent to the roundabout of the A47 Nechells Parkway and the A4540 Dartmouth Middleway. Come off the M6 at Jct 6 and go south on to the A38 Aston Expressway towards Birmingham and follow the brown signs.
Tel: 0121-202-2222.
www.thinktank.ac
E.mail: findout@thinktank.ac
Opening times: 10.00 - 17.00 every day except 24 - 26th December.
Admission: £9.00 + Concessions.

Foden	848	DD 4894	1904	-	Wagon
Ruston Proctor	18188	-	1894	-	Portable
Savage	739	-	1898	-	Organ Engine

Birmingham, West Midlands Fire & Rescue Service
West Midlands Fire Service Headquarters
99 Vauxhall Road
Birmingham
B7 4HW
Directions: situated in the City Centre of Birmingham, just east of the A4540. Proceed along that road until you reach Curzon Circle roundabout, turn east into Vauxhall Road, the entrance to the fire station is just along here on the right.
Opening times: Check web-site.
Admission: Check web-site.

| Shand Mason | 1592 | - | 1901 | Florian | FE |

Note: *This engine is not currently on public display due to a pending move to a new brigade museum.*

Blackgang Chine, Nr Ventnor Isle of Wight
Blackgang Chine Fantasy Park

Nr Ventnor
Isle of Wight
PO38 2HN
Directions: Located to the south of the island just off the A3055 road in Blackgang itself.
Tel: 01983-730052.
www.blackgangchine.com
E.mail: info@blackgangchine.com
Opening times: 30th Mar - 1st Nov: Times vary so consult web-site for more information.
Admission: £9.50 with concessions.

| Ruston Proctor | 51457 | - | 1916 | - | Portable |

Blaenafon, Big Pit National Coal Museum
Blaenafon
Torfaen
NP4 9XP
Directions: Located just south of the A465 Heads of the Valley Road, at the convergence of the B4248 and B4246. At that junction turn right into North Street, take the next left into Estate Road, take first left into an industrial estate continue over the railway bridge and as the road swings left take first left and follow the signs to the museum.
Tel: 01495-790311.
www.museumwales.ac.uk/en/bigpit/
E.mail: See Web-site
Opening times: Daily 09.30 - 17.00, but check for Dec & Jan.
Admission: Free.

| Marshall | UP26 | - | - | - | Portable |

Bradford Industrial Museum
Moorside Road
Eccleshill
Bradford
BD2 3HP
Directions: The museum is located 1¾ miles north east of the city. Take the A658 Harrogate Road pass Wellington Street on your left and at the next cross-roads turn right into Leeds Road, after 300 mtrs turn left into Musgrave Road at the end of which turn left into Moorside Road, the museum is on the right about 300mtrs further on.
Tel: 01274-435900.
www.bradfordmuseums.org
E.mail:industrial.museum@bradford.gov.uk

Opening times: Tues - Sat: 10.00 - 17.00. Sun: 12.00 - 17.00. Open also on bank holidays. Closed Good Friday & 25/26th Dec.
Admission: Free.

| Wallis & Steevens 7986 | OT 8207 | 1928 | - | RR |

Brading, Isle of Wight, 'The Experience'
High Street
Brading
Isle of Wight
PO36 0DQ
Directions: Located on the main A3055 between Ryde and Sandown, in the centre of Brading.
Tel: 01903-407286.
www.bradingtheexperience.co.uk
E.mail: info@bradingtheexperience.co.uk
Opening Times; Easter - October: 10.00 - 17.00. Nov - Easter: check web-site.
Admission: £7.25 with concessions.

Aveling & Porter	10626	DL 3128	1923	Borough of Ryde	RR
Fowler	7769	HC 2431	1898	Lettuce	TE
Garrett	33818	BJ 5323	1920	Pride of Wight	Shwmn
Marshall	87087	PO 7927	1933	Dorothy	TE
Wallis & Steevens	7650	DL 2864	1922	James the First	RR Conv.

Note: This location has closed at the end of 2009; there is no information about any re-opening date.

Brentwood, Old MacDonald's Farm
Weald Road,
Brentwood
Essex
CM14 5AY
Directions: From M25 come off at Jct 28 and take the A1028 to Brentwood, turn left at first traffic lights into Wigley Bush Lane, after ½ mile turn left at 'T' Jct into Weald Road, 2 miles on farm is on left at Jct with Chequers Road.
Tel: 01277-375177.
www.omdfarm.co.uk
E.mail: info@oldmacdonaldsfarm.org.uk
Opening times: 1st April - 30th Sept: 10.00 - 17.00 daily. 1st Oct - 31st Mar: 10.00 - 16.00. Weekends only in January.
Admission: £10.50 with concessions. £5 after 3pm.

| Marshall | - | - | - | - | Portable |

Note: This engine can easily been seen from the M25.

Bridport, Highlands End Holiday Park
Highlands End Holiday Park
Eype
Nr Bridport
Dorset. DT6 6AR
Directions: Head for Bridport, on the A35 Honiton - Dorchester Road, The holiday park is just to the south of this road west of the town. Follow the brown tourist signs.
Tel: 01308-422139.
www.wdhl.co.uk
E.mail: holidays@wdhl.co.uk
Opening times: The site is open through the summer months.
Admission: The engine can be seen from a pathway passing the main building in which it is housed, so there is no need to gain admission to the Park. This pathway is accessible all year round.

Note: If you wish to gain access to the vehicle for photographic or other purposes, please enquire at reception.

| Merryweather | - | - | 1902 | Bridport | FB | FE |

Brighton, British Engineerium,
The Droveway
Hove. BN3 7QA
Directions: Located just to the north of Hove Greyhound Stadium. From the A27 come off at Jct. with A2038 and head into Brighton on King George VI Avenue, after ¾ mile bear left into Nevill Road, the A2023. After another ½ mile you come to a crossroads, the museum is located on the junction.
www.britishengineerium.com
E.mail: info@britishengineerium.com

Note: This location is temporarily closed for refurbishment and should open in 2010. Some of the above engines may have moved from this site, see web-site for details.

Burrell	3786	PB 9610	1918	Tiger	TR
Merryweather	8558/B785	-	1944	-	FP
RSJ	E409	-	1930	-	Portable
Shand Mason	-	-	1892	Bor. of Barnstable	FE

Bristol Industrial Museum
This location is closed for the foreseeable future.

Breamore House, near Fordingbridge is the setting for this shot: Burrell 4053, Reg: TD 8047 shares a new building with numerous tractors and other farm equipment as well as a Robey portable. Photo taken on 28th August 2008.

One of the more unlikely locations for a 'steamer', Farmer Parr's Animal World Museum at Fleetwood, houses this Burrell No. 3305 amongst a great array of other historic items. Captured on 2nd July 2005.

Although currently closed the Coors Brewery Museum at Burton-on-Trent is home to this Sentinel No. 1488 dating from 1917. Reg: AW 3407. Hopefully this will re-open in the not too distant future. Picture taken on 13th October 2007.

Following the closure of the old Birmingham Science Museum in Newhall Street, the new 'Thinktank' building is grand and futuristic, but photography of the exhibits here is difficult. Foden Wagon 848 of 1904 illustrates the point.

'John', a Fowler Plough at the Museum of Lincolnshire Life on16th August 2008. One of a pair; 16053/4. 'Michael', the other half of the pair is in currently in their store shed. They were built in 1925.

An excellently presented 'street scene' museum is to be seen at Milestones, Basingstoke. Here on 6th September 2009 Tasker No 1726 Reg: SR 1294 from 1917 looks well alongside the old gas lamp and lumber wagon.

Sentinel Wagons are rarely found in museums, but there are one or two on display such as this 1916 example, No.1286 at the Glasgow Museum of Transport. **Brian Sharp.**

Not all museum engines are exhibited in pristine condition. At Luton Stockwood Park museum this Aveling & Porter, No. 10471 named 'Dennis' shows what lies under the smooth painted boiler cladding. Photographed on 27th March 2009.

Brownhills, Chasewater Railway
Chasewater Country Park
Pool Road
Brownhills
Staffs. WS8 7NL
Directions: Located just north of Brownhills. From the A5 (Watling Street) brown tourist signs will direct to the site. The engine is at Brownhills West Station Museum.
Tel:01543-452673.
www.chasewaterrailway.co.uk
E.mail: info@chasewaterrailway.co.uuk
Opening times: See Web-site.
Admission: See below.

Note: This is an operational steam railway and prices vary according to what you wish to do.

| Merryweather | 4716 | - | - | - | Ex Oswestry 2 wh. FP |

Burton-on-Trent, Coors Museum
This location expected to re-open in the near future.

| Sentinel | 1488 | AW 3407 | 1917 | Old No.1 | Wagon |

Burton-on-Trent, Staffordshire Fire & Rescue Educational Centre and Museum
Moor Street
Burton-on-Trent
Staffordshire
DE14 3SU
Directions: Located in the centre of the town near the railway station. From the A38 turn east into the town on the B5017, Shobna Road, at the cross-roads go straight on, this is the A5189; go over the railway bridge, at the roundabout turn first left into Anglesey Road, at the 'T' Jct turn right into Moor Street. The fire station and museum is along here.
Tel: 01785-898070.
www.staffordshirefire.gov.uk
E.mail: See web-site
Opening times: By prior appointment. Ring the above number to arrange.
Admission: Free, but a donation is welcomed.

| William Rose | - | - | 1899? | Georgina Rose | FE |

Note 1: This engine has been thought to be a Shand Mason for many years. A recent survey of it has proved that not to be the case. The engine can be seen from the roadside.

Bury, East Lancashire Railway
This location is temporarily closed for refurbishment

| Marshall | 71396 | - | 1919 | Doris | Portable |
| T. Green | 1978 | U 9647 | 1921 | Hilda | RR |

Caister Castle Museum
Castle Lane
Caister-on-Sea
Great Yarmouth
Norfolk
NR30 5SN
Directions: From Caister take the A149 Norwich road, at roundabout *(Jct with A1064)* take 1st left into Castle Lane, the museum is on right.
Tel: 01572-787649.
www.caistercastle.co.uk
E.mail: No e.mail address
Opening times: Mid May to last Friday in September; 10.00 - 16.30.
Admission: £9.00 with Concessions.

Note: This museum also houses a collection of cars including some steam powered.

| Marshall | 26660 | - | 1895 | - | Portable |

Calbourne Water Mill, Isle of Wight
Calbourne
Westover
Newport
Isle of Wight
PO30 4JN
Directions: From Newport head west on the B3323, after ¾ mile take the right fork on the B3401 Calbourne Road, continue through Calbourne, where the B3401 splits, take the left fork, the watermill is a right turn 200mtrs along.
Tel; 01983-531227.
www.calbournewatermill.co.uk
E.mail: thewatermill@hotmail.co.uk
Opening times: 26th Mar - 1st Nov: 10.00 - 17.00, 7 days a week.
Admission: £7.00 with Concessions.

| Tasker | 1235 | - | 1899 | - | Portable |

Cambridge Museum of Technology
Note: The Eddington & Steevenson Portable has left this location and is now in France. However there are a number of industrial steam engines on site.

Canterbury, Preston Services
Preston
Canterbury
Kent
CT3 1DH
Directions: Located south of the A28 Canterbury - Margate Road. Turn right on to the A253 heading for Ramsgate, after 1¼ miles turn right again on to the B2046 to Wingham, pass over the River Stour through East Stourmouth and on to the village of Preston, turn right into Court Lane, Preston Services is at the end of this lane by a church.
Tel: 01227-722502.
www.prestonservices.co.uk
E.mail: mlb@prestonservices.co.uk
Opening times: See web site for special events.
At other times strictly by appointment only.
Admission: Varies according to event.

Note: This location has an ever changing list of engines as they are constantly bought and sold. There are however a considerable number of portables (often 50-60) here as well as numerous traction engines of varying types and makes. Your visit will be well rewarded.

Cardiff, Welsh Ind & Maritime Museum
This location is closed for the foreseeable future.

Aveling & Barford	AH 412	ECT 452	1948	-	RR
Aveling & Porter	8906	DE 5880	1918	Morganydd	RR
RSJ	31136	DM 3048	1920	-	TE

Carrickfergus Museum & Civic Centre
11 Antrim Street
Carrickfergus
Co. Antrim
BT38 7DG
Directions: Located just off the Main A2 Larne - Belfast road in the centre of Carrickfergus. From the south come into town on the A2 and turn left into Castle Street continue into High Street and turn left into Antrim Street. From the north, turn right into Joymount and turn left at 'T' Jct after which the road bears right into Antrim Street.
Tel: 028-9335-8000.

www.carrickfergus.org
E.mail: info@carrickfergus.org
Opening times: Apr - Sept: Mon - Sat: 10.00 - 18.00. Sun: 13.00 - 18.00. Oct - Mar: Mon - Sat: 10.00 - 17.00. Sun: 13.00 - 17.00. Closed Bank hols.
Admission: Free.

| Shand Mason | - | - | 1908 | Carrickfergus F.B. | FE |

Chatham Dockyard
The Historic Dockyard
Chatham
Kent
ME4 4TZ
Directions: From A2/M2 Jct 1 follow Brown tourist signs along the A289 and Medway Tunnel which will take you straight there.
Tel: 01634-823807.
www.chdt.org.uk
E.mail: info@chdt.org.uk
Opening times: Daily from 14th Feb - 1st Nov: 10.00 - 16.00 (18.00 from 29th Mar - 24th Oct). Weekends only in Nov. Closed Dec & Jan.
Admission: Admission varies according to event.

Note 1: This collection of engines may change from time to time, not all those listed are permanently based at Chatham. Access to them may only possible during official open days, additional engines are often present. See web-site for dates and other details.

Note 2: The portable steam raiser was built at Chatham and is stored in the yard.

Albaret	946	-	1923	-	RR
Aveling Barford	AH162	JXH 174	1948	Omega	RR
Aveling & Porter	5156	TU 874	1902	Wild Rose	RR
Aveling & Porter	8097	FX 7014	1913	Moby Dick	RR
Aveling & Porter	10271	ME 2103	1922	King George	RR
Aveling & Porter	10399	BH 9624	1922	John Hampden	RR
Aveling & Porter	14044	DW 7088	1930	Rhoda	RR
Aveling & Porter	14072	DV 7077	1930	-	RR
Aveling & Porter	14073	FV 1326	1930	Andrea	RR
Aveling & Porter	3567	-	1895	Sydenham	Rail loco
Chatham	-	-	-	-	Steam raiser
Fowler	15698	BW 6179	1922	Lugtrout Lion	RR
Fowler	15732	PT 832	1923	Sir Douglas	Tr
Marshall	20146	-	1891	-	Portable
Sentinel	9208	BYL 485	1935	-	Wagon
Shand Mason	-	-	-	Kent Fire Brigade	FE
Wallis & Steevens	8033	OU 5185	1930	Phoebe	RR

Chester, Cheshire Fire & Rescue Service
Chester Fire Station
St Anne Street
Chester
CH1 2HP
Directions: Located just to the north-west of the St Oswald's Way ring road around the city centre, between the A5118 and the A56. St Anne Street is a turning off this ring road.
Tel: 07776-297784.
www.cheshirefire.gov.uk
E.mail: john.edwards@cheshirefire.gov.uk
Opening times: By appointment only.
Admission: Free but donations welcomed.

Note: This engine is not on public display, The Cheshire Fire Service does however permit visitors by prior arrangement, also the engine can often be seen at traction engine rallies and other allied events in the area.

| Shand Mason | - | - | 1880? | City of Chester | FE |

Chesterfield, Barrow Hill Roundhouse Railway Centre
Campbell Drive
Barrow Hill
Chesterfield
Derbys.
S43 2PR
Directions: From M1 come off at Jct 29 or 30 on to the A619 heading west, you will then be able to follow the brown tourist signs for the centre.
Tel: 01246-472450.
www.barrowhill.org
E.mail: projectman@barrowhill.org.uk
Opening Times: Sat & Sun: 10.00 - 16.00.
See web-site for special events
Admission: Charges vary according to event. Donations on non-event days

Note: This is a steam railway centre with just this one resident roller on display.

| Aveling & Porter | 8778 | E 5359 | 1916 | Old Faithful | RR |

Chinnor & Princes Risborough Railway
Chinnor Station
Station Road
Chinnor
Oxon

OX39 4ER
Directions: From M40 come off at Jct 6 on to B4009 heading for Chinnor, then follow the brown tourist signs to station.
Tel: 01944-353535
www.chinnorraiilay.co.uk
E.mail: enquiries@chinnorrailway.co.uk
Opening times: See timetable on web-site
Admission: See web-site.

| Aveling & Porter | 9449 | - | 1926 | Blue Circle | Rail loco |

Chippenham, Lackham Museum
Lackham Museum of Agriculture and Rural Life Trust
Wiltshire College
Lackham
Chippenham
SN15 2NY
Directions: Located 3 miles south of Chippenham. From the A350 heading south go straight over the roundabout with the A4 to the next roundabout, again straight on, on the left is the entrance to the College and Museum.
Tel: 01249-466841.
www.lackhamcountrypark.co.uk
E.mail: enquiries@lackham.co.uk
Opening times:
Currently shut for re-organisation, please check web site.
Admission: See web-site.

| Brown & May | 6226 | - | 1899 | - | Portable |

Coalville, Snibston Discovery Park
Ashby Road,
Coalville,
Leics.
LE67 3LN
Directions: From M1 go west off Jct 22 on to A511 Buxton Road, at 1st roundabout turn right staying on the A511, straight on at next into Bardon Road, continue into London Road, Hotel Street and past Memorial Square into Ashby Road, museum is on left along here.
Tel: 01530-278444.
www.Leics.gov.uk/museums/snibston
E.mail: snibston@leics.gov.uk
Opening times: Apr - Sept; 10.00 - 17.00 daily. Oct - Mar: Mon - Fri: 10.00 - 15.00. Sat & Sun: 10.00 - 17.00. Closed Christmas and 1 week in January.
Admission: £6.75 with concessions.

| Marshall | 68823 | BE 3044 | 1915 | Jinglin' Geordie | TE |
| Shand Mason | - | - | - | Shepshed | FE |

Coatbridge, Summerlee Heritage Park
Heritage Way
Coatbridge
ML5 1QD
Directions: From the M8 proceed on to the A8 and then the A725 into Coatbridge town centre. At roundabout take first exit on to the A89, at the next roundabout take 2nd left go under the railway and turn immediately right into Heritage Way. Follow brown tourist signs. The museum is on the right.
Tel: 01236-638450.
www.northlan.gov.uk
E.mail: generalenquiries@northlan.gov.uk
Opening Times: Daily: 10.00 - 17.00 (16.00 Nov - Mar).
Admission: Free.

Fowler	17251	DS 7206	1928	-	RR
Sentinel	5676	EC 5927	1924	Big Tam	Wagon
There is also a 1985 'Houston' built freelance engine here named 'Tigger'						

Denby Dale, Nortonthorpe Mills
This engine is located within the Mill factory and is not currently on public view.

| Savage No.3 | - | - | 1895? | - | Organ Engine |

Dereham, Gressenhall Farm & Workhouse
Gressenhall
Dereham
Norfolk
NR20 4DR
Directions: Located 3 miles northwest of Dereham on B1146, follow brown tourist signs,
Tel:01362-860563.
www.museum,norfolk.gov.uk
E.mail: gressenhall.museum@norfolk.gov.uk
Opening times:
22nd March - 1st Nov; 10.00 - 17.00.
Admission:
£8.10 with concessions.

| Garrett | - | - | - | - | Portable |
| Tidman | - | - | 1885 | - | Portable |

Didcot Railway Centre
Didcot Station
Oxfordshire
OX11 7NJ
Directions: From M4, take Jct 13, the museum is signposted. Follow these into Didcot and to the railway station car park. The entrance to the museum is through an underpass below the station platforms.
Tel: 01235-817200.
www.didcotrailwaycentre.org.uk
E.mail: info@didcotrailwaycentre.org.uk
Opening times: Sat & Sun: 10.30 - 17.00 plus other times, check web-site.
Admission: £5 on non-steaming days. £8 when in steam. Other prices for special events, check web-site for details.

Merryweather	1541	-	1890	Blenheim	FE

Note: This engine was previously located at Shipston-on-Stour fire station

Diss, Bressingham Steam Museum Trust
Thetford Road
Nr Diss
Norfolk
IP22 2AB
Directions: From Diss take the A1066 west towards Thetford, the museum and gardens are 2½ miles along this road. The site is well sign-posted.
Tel: 01379-686900.
www.bressingham.co.uk
E.mail: info@bressingham.co.uk
Opening times: Mar, Apr, May, Sept, Oct: 10.30 - 17.00
Jun - Aug: 10.30 - 17.30.
Admission: £9.50 with concessions.

Burrell	2363	-	1901	-	Portable
Burrell	3112	CF 3440	1909	Bertha	Scc TE
Burrell	3962	PW 1714	1923	Boxer	RR
Burrell	3993	CF 5646	1924	Buster	RR
Foden	13708	VF 8862	1930	Boadicea	Wagon
Foster	2821	BE 7448	1903	Beryl	TE
Fowler	6188	MA 8528	1890	Beulah	TE
Garrett	34641	CF 5913	1924	Bunty	Tr
Merryweather	-	-	1914	-	FP
Merryweather	3702	-	-	-	FE
Robey	42520	FE 7632	1925	Barkus	Tandem RR
Shand Mason	-	-	1895	J.J. Coleman	Static engine
Tidman	-	-	1891	-	OE - Carousel
Tidman	-	-	1891	-	CE - Carousel
Youngs	-	-	1910	-	Portable

Doncaster, Cusworth Hall and Park
Cusworth Hall and Park
Cusworth Lane,
Doncaster
DN5 7TU
Directions; Located 2 miles north of Doncaster off A638 Wakefield Road, it is signposted from both the A1 and the A638.
Tel: 01302-782342.
www.doncaster.gov.uk/museums
E.mail: museum@doncaster.co.uk
Opening times: All year; Mon - Fri: 10.30 - 17.00. Sat & Sun: 13.00 - 17.00. Special exhibitions take place through the year, check web site for details.
Admission: Free, but there is a car park charge.

Merryweather	2428	-	1940?	Manvers Main Colliery Fire Pump Wath on Dearne 117

Dorchester, Greys Bridge Park
Greys Bridge
Fordington
Dorchester
Directions: Located in a recreation park near Greys Bridge, in Fordington, a suburb of Dorchester. Just to the east of the town near the A35 trunk road and River Frome.
Admission: Access is free to view this engine.

Aveling & Porter	10317	FX 9412	1922	-	RR

Dover Transport Museum
Willingdon Road
Whitfield
Dover
Kent
CT16 2HQ
Directions: Located south of the A2 in Dover. At the roundabout with the A256 go straight on into Honeywood Road, at next roundabout turn right into Menzies Road, the museum is on the right about 100mtrs further on.
Tel: 01304-822409.
www.dovertransportmuseum.org.uk
E.mail: see web-site.
Opening times: Easter to mid Sept: Wed - Fri: 13.30 - 16.30. Sun: 10.30 - 17.00 and Bank Hols. Winter: Sundays only 11.00 - 15.00.
Admission: £4.00 with concessions.

| Aveling & Porter 11055 | NY 6931 | 1924 | - | RR |

Draycott-in-the-Clay, Klondyke Mill
Klondyke Mill,
Draycott-in-the-clay
Staffs
DE6 5GZ
Directions: Located just south of the A50 between Uttoxeter and Tutbury. From A50 go south on A515 heading for Lichfield, after a level crossing keep on for about 3 miles, as you enter Draycott look for a white gate on your left, the engines are in the field behind.
Tel: 01538-360042.
www. nstec.co.uk
E.mail: info@nstec.co.uk
Opening times: See web-site.
Admission: See web-site.
Note1: Although some of the engines can be seen from the road-side, this is the home base of the North Staffordshire Traction Engine Club and the site is open on special events days only.

Note 2: Due to the nature of this location engines may change around. Please check web-site before visiting.

Burrell	2706	TB 2845	1904	Admiral Togo	TE
Fowler	15337	DO 1928	1919	-	Plg
Fowler	15787	EP 2398	1924	Cynorthwywr	Shwmns
Fowler	18507	SM 8832	1931	Morning Star	RR
Marshall	46699	-	1907	-	Portable
Marshall	59393	-	1912	-	Portable
Marshall	81427	PY 6079	1926	Anne	RR
RSJ	14329	-	1902	-	Portable
T. Green	2054	RF 3309	1927	-	RR

Dublin, The National Transport Museum, Howth Castle
Heritage Depot, Demesne,
Howth,
Dublin 13
Directions: Situated east of Dublin. Head for Howth on the R105 which continues as Howth Road, in Howth turn right into Howth Castle Road, the museum is on the right along here.
Tel: 01832-0427.
www.nationaltransportmuseum.org/
E.mail: info@nationaltransportmuseum.org
Opening Times: Sept - May: Sat, Sun & Bk Hols: 14.00 - 17.00. Jun - Aug: Mon - Sat: 10.00 - 17.00.

Admission: £3.00 with concessions.

Aveling & Porter	10617	IK 7224	1923	No.7	RR*
Merryweather	-	-	1889	GNR Dundalk Works	FE

Note: **Not on public view at present*

Dublin, Straffan Steam Museum
Lodge Park
Straffan
Co. Kildare
Directions: Located 16 miles west of Dublin, just north of route N7. Turn right at village of Kill; continue northwards for 4 miles, museum is on the right.
Tel:(353) (1) 6273155.
www.steam-museum.com/steam.html
E.mail: info@steam-museum.ie
Opening times:
Jun - Aug: Wed & Sun: 14.00 - 18.00 + bank hols. Other times by appointment.
Admission: 7.50 Euros.

Marshall	31910	-	1899	-	Portable

Dunfermline, Scottish Vintage Bus Museum
M90 Commerce Park
Lathalmond
Dunfermline
Fife
KY12 0SJ
Directions: Located 4 miles to the north of Dunfermline. From M90 turn west at Jct 4 on the B914 continue through Lassodie, after approx 1 mile turn left on the B915, museum is just along here on the right.
Tel:01383-623380.
www.busweb.co.uk/svbm/
E.mail: See Web-site.
Opening times:
Easter Sun - 4th Oct: 12.30 -17.00 and Special events.
Admission: Prices vary, see web site for details and special events.

Marshall	85601	SC7488	1930	-	RR

Dunrobin Castle
Dunrobin
Golspie
Sutherland

KW10 6SF
Directions: Located on the A9 ½ mile north of Golspie. Follow brown tourist signs.
Tel:01408-633177.
www.dunrobincastle.co.uk
E.mail: info@dunrobincastle.co.uk
Opening times:
1st April - 15th Oct: Mon - Sat: 10.30 -16.00.
Sun: 12.00 - 16.30.
Admission:
£8.00 with concessions.

| Shand Mason | - | - | 1903 | Dunrobin Castle | FE |

Note: This engine is located in the cafe!

Edinburgh, Lothian and Borders Fire and Rescue Service
Museum of Fire
Lauriston Place
Edinburgh
EH3 9DE
Directions: Situated south of the city centre near the junction of the A700 and A702. At this junction turn east on to Laursiton Place, the museum is about a ½ mile along at the junction with Lady Lawson Street.
Tel: 0131-659-7331 or free-phone 0800-169-0320.
www. lothian.fire-uk.org/museum
E.mail: enquiries@lbfire.org.uk
Opening times: Mon - Fri: 09.00 - 16.00. Closed for 2 weeks in August and over Christmas.
Admission: Free, but a donation is appreciated.
Note: Prior appointment is essential.

| Shand Mason | - | - | 1873 | Welbeck | FE |

Edinburgh, National Museums Collections Centre
Granton Centre
242 West Granton Road
Edinburgh
EH5 1JA
Directions: Located to the north of the City, from the A902 Ferry Road head north on Granton Road after a left bend in this road it continues into West Granton Road, the museum is along here.
Tel; 0131-247-4470.
www.nms.ac.uk/museumscollectioncentre

E.mail: info@nms.ac.uk
Opening Times: See above.
Appointments are possible by ringing 0131-247-4274 or;
E.mail: m.kerr@nms.ac.uk
Admission: Free
Note: This museum currently undergoing refurbishment and the engines are not on public display. Check web-site for further information about re-opening date.

| Marshall | 47731 | MS 3081 | 1907 | Sir Hector | TE |
| Tuxford | 1234 | - | 1886 | - | Portable |

Edinburgh, Royal Museum of Scotland
Chambers Street
Edinburgh
EH1 1JF
Directions: Located just south of Waverley Station to the west of the A47 in Chambers Street.
Tel:0131-247-4422
www.explore-edinburgh.com/museum
E.mail: info@nms.ac.uk
Opening Times:
Mon - Sat: 10.00 - 17.00 (Tues 'til 20.00) Sun: 12.00 - 17.00.
Admission: Free.

Note: This location is now closed, it should re-open in 2011, the engines are currently stored at the Granton Centre (see above)

Ely Fire Station
Egremont Street
Ely
Cambs. CB6 1AE
Directions: From the A10 Ely By-pass north of the City turn on to the B1411 Downham Road, this leads into Egremont Street, the fire station is on the left.
Tel: 01353-662223.
www.cambsfire.gov.uk
E.mail: go to web-site.
Opening Times: By appointment only.
Admission: Free, but a donation is appreciated.

Note: This engine is not currently on public display.

| Shand Mason | - | - | 1912 | Ely Fire Station | FE |

Enfield, Whitewebbs Museum of Transport
Whitewebbs Road

Enfield
Middlx
EN2 9HW
Directions; From M25 take Jct 25 and turn right on to A10 heading for London. Turn right into Bullsmore Road, at 'T' Jct turn right then road bears left into Whitewebbs Road, museum in on your left.
Tel: 020-8367-1898.
www.whitewebbsmuseum.co.uk
E.mail: whitewebbsmuseum@aol.com
Opening times: Tuesdays: 10.00 - 16.00 & last Sunday of each month 10.00 - 16.00.
Admission; £3.00.

Shand Mason	-	-	1896	Hazelmere	FE

Exmouth, Sandy Bay World of Country Life
Sandy Bay
West Down Lane
Littleham, Exmouth
Devon
EX8 5BY
Directions: This is located 10 miles to the south of the M5, come off at Jct 30 and follow A376 sign-posted 'Exmouth' & 'Sandy Bay'. Follow brown tourist signs.
Tel: 01395-274533.
www.worldofcountrylife.co.uk
E.mail: info@worldofcountrylife.co.uk
Opening times:
1st April -1st November: 10.00 - 17.00.
Admission:
£9.85 with concessions.

Burrell	2877	HT 3163	1907	His Majesty	Shwmn
Burrell	3711	TA 3067	1916	King 'the Belgians	Shwmn
Burrell	3884	XH 5728	1921	Gladiator	Shwmn
Fowler	9971	HO 5655	1904	Candyfloss	Class D2
Merryweather	-	-	1885	-	FE
Robey	43265	FE 9350	1927	Sir Charles	Exp. Tractor
Waterloo	1666	-	1913	-	TE

Fakenham, Thursford Collection,
Thursford
Fakenham
Norfolk
NR21 0AS

Directions: Situated 1 mile north of the A148 between Fakenham and Holt. Follow brown tourist signs.
Tel: 01328-878477.
www.thursford.com
E.mail: admin@thursfordcollection.co.uk
Opening times: 5th Apr - 27th Sept: 12.00 - 17.00, closed Sats. Except Easter Sat. For other opening times and special events see web-site.
Admission; £8.00 with concessions. See web-site for special events prices.

Note: Engines in this collection are stored or displayed in different locations around the site. Thus it may not always be possible to see the whole collection on any one visit.

Aveling & Porter	8169	KT 998	1913	-	TE
Aveling & Porter	8178	PU 700	1914	-	Conv
Aveling & Porter	8200	AH 0778	1914	-	RR
Aveling & Porter	8890	DO 1943	1918	Field Marshall Haig	Plg
Aveling & Porter	8891	DO 1944	1918	General Byng	Plg
Aveling & Porter	9010	AP 9235	1919	Jimmy	TE
Aveling & Porter	9036	BP 6065	1919	-	Conv
Aveling & Porter	9149	KE 2202	1920	-	RL
Aveling & Porter	10003	AH 6130	1921	-	Conv
Aveling & Porter	10324	NO 5898	1922	-	RR
Aveling & Porter	10341	PW 623	1922	-	Conv
Aveling & Porter	10342	NO 5896	1922	-	RR
Aveling & Porter	10345	NO 5893	1922	-	RR
Aveling & Porter	10347	NO 5891	1922	-	RR
Aveling & Porter	10415	ME 5770	1922	-	RR
Aveling & Porter	10437	OS 1314	1925	-	Conv
Aveling & Porter	10456	NO 7239	1922	-	RR
Aveling & Porter	10675	PR 1070	1923	-	RR
Aveling & Porter	10755	PD 1510	1924	-	RR
Aveling & Porter	10780	PR 1392	1923	-	RR
Aveling & Porter	10785	PD 7738	1923	-	RR
Aveling & Porter	11205	MO 5549	1925	-	RR
Aveling & Porter	11322	PT 6445	1925	Castra	RR
Aveling & Porter	11454	PE 9234	1926	-	RR
Aveling & Porter	11822	UF 1934	1927	-	RR
Aveling & Porter	11918	YT 4531	1927	-	RR
Aveling & Porter	11980	DY 4878	1927	Bert	RR
Aveling & Porter	12181	DW 6156	1928	Cheryl	RR
Aveling & Porter	12186	PK 2684	1928	Annie Laurie	TR
Aveling & Porter	12205	VF 4393	1928	-	RR
Aveling & Porter	14008	VG 2269	1929	-	RR
Aveling & Porter	14137	TM 9357	1931	-	RR
Aveling & Porter	14163	MJ 4597	1934	-	RR
Burrell	2780	CL 4300	1905	King Edward VII	Shwmn
Burrell	3075	CL 4301	1909	Alexandra	Shwmn
Burrell	3200	CL 4296	1910	Unity	Shwmn
Burrell	3827	CL 4299	1920	Victory	Shwmn
Burrell	3850	AH 0775	1920	-	Conv
Burrell	4045	PW 8878	1926	-	TE

Clayton & Shutt	48308	FE 2754	1919	-	Wagon
Clayton & Shutt	49105	TL 555	1924	-	TE
Foden	13358	DF 8187	1929	Victor	TR
Garrett	33902	BJ 5340	1920	Medina	Shwmn
Garrett	34187	AH 9623	1922	Kathleen	TE
Marshall	79108	PW 5042	1925	Poppyland	RR
Ruston Proctor	39872	RT 1487	1910	-	TE
Savage	418	-	1887	-	OE
Savage	421	-	1887	-	CE
Savage	449	-	1888	-	OE
Savage	740	-	1898	-	CE
Savage	762	-	1900	Enterprise	CE
Savage	763	-	1900	-	OE

Fleetwood, Farmer Parrs Animal World

Wyrefield Farm
Rossall Lane
Fleetwood. FY7 8JP
Directions: Situated on the Wyre Peninsula between Thornton and Fleetwood. From M55, come off at Jct 3 and head north on A585 Fleetwood Road for several miles until you arrive at the roundabout with the B5412, go straight on, after 1½ miles, at next roundabout take the third exit almost turning back on yourself, this is Fleetwood Road North, the site is a few yards along here on your right.
Tel: 01253-874389.
www.farmerparrs.com
E.mail: enquiries@farmerparrs.com
Opening times: Daily 10.00 - 17.00.
Admission: £4.50 with concessions.

| Burrell | 3305 | TB 3722 | 1911 | - | RR |

Fochabers, Christies Garden Centre

Christies (Fochabers) Ltd
The Nurseries
Fochabers
Morayshire. IV32 7PF
Directions: In Fochabers itself, at the Junction of the A98 and A96 between Keith and Elgin.
www.christie-elite.co.uk
E.mail: christies@btinternet.com
Opening times: All year round.
Admission: Free.
Note: *The engine is on display by the main entrance*

| Robey | 53445 | - | 1955 | - | Portable steriliser |

Fordingbridge, Breamore House
Breamore
Fordingbridge
Hants. SP6 2DF
Directions: Situated west of the main Ringwood - Salisbury A338 road just north of the village of Breamore. Follow brown tourist signs along a narrow country lane, the museum is on the right about 1 mile on.
Tel: 01725-512858.
www.breamorehouse.com
E.mail: breamore@btinternet.com
Opening times: Various, check web-site.
Admission: £8 with concessions.

Burrell	4053	TD 8047	1926	The Dreadnought	DCC TE
Robey	40062	-	1921	Ruby	Portable
Shand Mason	-	-	1904	Hinton Admiral	Not on display

Forres, Dallas Dhu Historic Distillery
Mannachie Road
Forres
Morayshire
IV36 2RR
Directions: Located just south of the town of Forres. From the A96, turn south in the town and head for Grantown-on-Spey on the A940. After the road bears right look for a left turn which is Mannachie Road follow this for about 1½ miles, the distillery is sign-posted and on the left along here.
Tel: 01309-676548.
www.historic-scotland.gov.uk
E.mail: See web-site.
Opening times: Daily: 1st April - 30th Sept: 09.30 - 17.30. 1st Oct - 30th March: 09.30 - 16.30. Last admission; 30 minutes before closing time.
Admission: £5.20 + Concessions.

| Shand Mason | - | - | 1895? | Glenlossie | FE |

Glasgow Museum of Transport
1 Bunhouse Street
Glasgow
G3 8DP
Directions: From M8 Jct 19 head west on to the A814 Stobcross Street, past Anderston railway station, come off at 1st Jct down under the expressway and head north on Finnieston Street, at 'T' Jct with Argyll Street turn left and continue about 1 mile to just before the river and turn left into Bunhouse Road.
Tel:0141-287-2615.

www.glasgowmuseums.com
E.mail: museums@glasgow.org
Opening times: All year except Xmas and new year, for times check web-site.
Admission: Free.
Note; This museum will close in 2011 to move to a new location

Merryweather	1818	-	-	John Brown & Co.	Greenwich model
Ruston Hornsby	113812	MS 3273	1920	Pride of Endrick	TE
Sentinel	1286	AW 2964	1916	-	Wagon

Gloucester City Museum & Art Gallery
Brunswick Road
Gloucester
GL1 1HP
Directions: From M5 come off at Jct 11A, go west on A417, pass over three roundabouts, at the fourth turn right on to B4073. Stay on this to 'T' Jct and turn left there on to A430 Trier Way, follow on to next 'T' Jct, turn right on to A4301 Southgate Street. Take 4th right along here into Parliament Street which leads to Brunswick Road. The museum is on the left.
Tel:01452-396868.
www.gloucester.gov.uk/museums
E.mail: city.museum@gloucester.gov.uk
Opening times: Tues - Sat 10.00 - 17.00.
Admission: Free.

| Trotter | - | - | 1933 | - | Vertical boiler RR |

Greenock, Strathclyde Fire Preservation Group
Wallace Place
Greenock
PA15 1JB
Directions: Located at the main roundabout in Greenock where the A8 and A78 meet. From Glasgow on the A8 turn left at roundabout, then first left into Wallace Place, There is a small car park on the left and the museum is in front of you. From A78 go right round roundabout back on to the A78 and turn left into Wallace Place.
Tel: See Web-site.
www.strathclydefire.org.uk
E.mail: See web-site.
Opening times: This museum is currently being commissioned and will be open later in the year, see web-site for further details.
Admission: To be confirmed.

| Merryweather | - | - | 1900? | Victoria/Oxford | FE |

Harlington, Beds, Poplars Garden Centre
Harlington Road
Toddington
Dunstable
Beds
LU5 6HE
Directions: The garden centre is located adjacent to the M1, turn off at Jct 12 heading for Bedford on the A5120, about ½ mile on Poplars is on the right. You will see the engine as you approach, turn right in to the car park.
Tel: 01525-872017.
www.poplars.co.uk
E.mail: enquiries@poplars.co.uk
Opening times: not applicable.
Admission: not applicable.

Note: This is a commercial garden centre and not a museum, the engine is right by the roadside and easily accessible.

Robey	-	-	1950	-	Steam producer

Helston, Flambards, 'The Experience'
Helston
Cornwall
TR13 0QA
Directions: Head for Helston on the A394, the park is on the Helston side of Culdrose airfield just off that road. Follow the brown tourist signs.
Tel: 01326-573404 or: Info-line 0845-6018684.
www.flambards.co.uk
E.mail: info@flambards.co.uk
Opening times: Times vary so please check web-site.
Admission: £16.50.
Note: The engine is in the Victorian Village complex.

Shand Mason	-	-	-	Cliveden	FE

Hollycombe, Steam in the Country
Iron Hill
Liphook
Hants
GU30 7LP
Directions: Hollycombe is 2 miles from the A3 between Guildford and Petersfield. Leave the A3 at the Liphook turn and follow the brown tourist signs.
Tel: 01428-724900.
www.hollycombe.co.uk

E.mail: info@hollycombe.co.uk
Opening times: Suns and Bank Hols 10th April - 25th October. Daily 23 - 29th May & 4th - 31st August. Check web site for special events.
Admission: £11 with concessions.

Aveling & Porter	8653	KE 6455	1912	Jo Ann	RR
Aveling & Porter	10050	XM 4037	1921	David	RR
Aveling & Porter	10143	FX 9683	1922	Jasper	RR
Brouhot	-	-	1896	Bernadette	Portable
Brown & May	6691	-	1901	-	Portable
Burrell	1876	AH 5366	1895	Emperor	Shwmn
Burrell	3356	-	1912	Highflyer	Steam yachts
Burrell	3545	AH 0173	1914	Little Topper	Tractor
Burrell	3815	BP 5757	1919	Sunset No.2	Tractor
Clayton & Shutt.	44140	-	1911	-	Portable
Clayton & Shutt.	50010	-	1926	Eileen	Portable
Fowler	14383	NM 1284	1917	Prince	Plg
Garrett	33348	SZ 100	1918	Leiston Town	Bioscope
John Allen	67	BW 4613	1913	-	Plg
Mann	1260	U 4298	1917	Old King Cole	Tractor
Merryweather	2743	-	1908	Colstoun house	FE
Robey	33810	-	1915	-	Portable
Ruston Proctor	30656	-	1906	Big John	Portable
Savage	772	-	1900	Little Jim	Centre Engine
Savage	869	-	1915	Bottoms Pride	Centre Engine
Savage	900	-	1923	-	Organ Engine
Tidman	-	-	1881	-	Centre Engine
Tidman	-	-	1881	-	Organ Engine
Walker	-	-	1910	-	Centre Engine
Wallis & Steevens	8023	CG 571	1932	Christopher	Simplicity RR

Ipswich, Suffolk Fire Service
Colchester Road
Ipswich
IP4 4SS
Directions: The Fire Headquarters are located on the A1214 ring road to the north-east of the town. Between the B1077 and A1214, turn off for Woodbridge.
Tel: 01473-588942.
www.customerservices@csduk.com
E.mail; No e.mail
Opening times: By appointment only.
Admission: Not applicable.

Note: This is an operational fire station, not normally open to the public. The engine is however inside the building but can be viewed from the road.

Shand Mason	-	-	1911	Needham Market	FE

Ironbridge Gorge Museum, Blists Hill
Coach Road,
Coalbrookdale
Telford
TF8 7DQ
Directions: Blists Hill is 5 miles south of Telford. From the M54 take Jcts 4 or 6 and follow the brown tourist signs to Blists Hill.
Tel: 01952-884391.
www.ironbridge.org.uk
E.mail: tic@ironbridge.co.uk
Opening Times: 28th Mar - 8th Nov: 10.00 - 17.00 daily. 9th Nov - 19th Mar 2010: 10.00 - 16.00 daily. Closed 24/5th Dec. and 1st Jan.
Admission: £13.25 to Blists Hill site only. £19.95 to all sites.

Wallis & Steevens 2660	CJ 4816	1903	Billy	RR

Jersey, Pallot Steam, Motor & Gen. Museum
Rue de Bechet
Trinity
Channel Islands
Directions: Rue de Bechet is located between the main A9 St John's Road and the A8 Trinity Road in the town.
Tel: 01534-865307.
www.pallotmuseum.co.uk
E.mail: info@pallotmuseum.co.uk
Opening Times: Mon - Sat: 1st April - 31st Oct.10.00 - 17.00.
Other times by appointment.
Admission: £5 with concessions.

Merlin & Cie	-	-	1924	-	Portable
Merlin & Cie	-	-	?	-	In build
Marshall	78539	J 17884	1922	-	RR
Marshall	81451	-	1926	-	RR
RSJ	15740	J 14457	1904	Dolly May	TE

Kendal, Levens Hall
Levens Hall
Kendal
Cumbria. LA8 0PD
Directions: From M6 take Jct 36, turn west on to A590 for about 4 miles to Sedgwick, Turn Left at roundabout staying on the A590 heading for Barrow. After half mile turn left on to A6, after approx 1 mile Levens Hall is on right.
Tel: 015395-60321.
www.levenshall.co.uk

E.mail: email@levenshall.fsnet.co.uk
Opening times: Sun - Thurs: April - 1st Week of Oct: 10.00 - 17.00.
Admission: Variable - see web-site.

| Foden | 11892 | DD 7916 | 1925 | - | Wagon |
| Fowler | 15116 | CB 3062 | 1920 | Bertha | Shwmn |

Kenilworth, Stoneleigh Abbey
Estate Office
Kenilworth
CV8 2LF
Directions: Located just east of Kenilworth off the A46. From the roundabout with the A452, turn east towards Leamington Spa. After a short distance turn left on to the B4115, go over a river bridge and continue to a red sandstone gatehouse on your right, turn left opposite that into car park, entrance to Abbey is through gatehouse.
Tel: 01926-858585.
www.stoneleighabbey.org
E.mail: enquire@stoneleighabbey.org
Opening times: Tues, Wed, Thurs, Sun & Bank hols: 10.30 - 16.00.
Admission: £6.50 with concessions.

| Shand Mason | - | - | 1890 | - | FP |
| Shand Mason | - | - | 1909 | Francis Dudley | FE |

Kew Bridge Steam Museum
Green Dragon Lane
Brentford
Middlx
TW8 0EN
Directions: Located on the north bank of the Thames on the corner of Kew Bridge Road and Green Dragon Lane. Look for tall Stand Pipe tower. Take tube to Gunnersbury then walk or bus 237 or 67. Alternatively from Kew Gardens tube station bus, 65, 237, 267, and 391.
Tel: 020-8568-4757.
www.kbsm.org
E.mail: info@kbsm.org
Opening times:
11.00 – 16.00, Tues - Sat & Bank holidays.
Admission: £9.50 with concessions.

Note: At present there are no engines permanently based here, but various locomotives do visit on special events days. Check web-site for further details.

Kidderminster, Hodgehill Nurseries Ltd
Birmingham Road
Kidderminster
DY10 3NR
Directions: Located on the west-bound side of the main A456 road between Kidderminster and Hagley.
Tel: 01562-822058.
www.hodgehill-ltd.co.uk
E.mail: hodgehill-ltd@btconnect.com
Opening times: Mon - Sat: 8.30 - 17.30, all year round.
Admission: Free.

Note: This is a commercial garden nursery; however the engine is parked by the road-side at the entrance and can easily be seen from the road.

Davey Paxman	UP9	-	-	-	Portable

Knutsford, Tatton Hall
Tatton Park
Knutsford
Cheshire
WA16 6QN
Directions: From M6 Jct19 turn east on to A556 Altrincham Road, after 2 miles turn right at cross-roads on to A50, continue for 1 mile then turn left on to A5034. After 400yds turn right on to a minor road to Ashley, Tatton Hall is on right.
Tel:01625-534400.
www.tattonpark.org.uk
E.mail: tatton@cheshire.gov.uk
Opening times: All year: 28th Mar - 4th Oct: 10.00 - 19.00. 5th Oct - 26th Mar: 11.00 - 17.00.
Admission: Various, according to which part you visit - See details on Web-site

Note: the engine is displayed in the New Barn in the stable-yard.

Shand Mason	-	-	1890	Tatton	FE

Launceston, Dingles Fairground Heritage Centre
Milford Lifton
Devon
PL16 0AT
Directions: From A30 head for Tavistock town centre, from the church take the Chillaton Road and follow the brown tourist signs for 12 miles to Dingles which is near the village of Milford Lifton.

Tel: 01566-783425.
www.fairground-heritage.org.uk
E.mail: richard@dinglesteam.co.uk
Opening times: 20th Mar - 3rd Nov: Thurs - Mon: 10.30 - 17.30.
Admission: £7.00 with concessions, also check web- site for special events.

Burrell	3509	G 6463	1913	Rajah	Shwmn
Burrell	3986	AF 9293	1924	-	RR
Burrell	3996	E 9597	1924	Conqueror	RL
Garrett	34193	-	1924	-	Portable

Launceston Steam Railway
St Thomas Road
Launceston
PL15 8DA
Directions: Located in Launceston itself just north of the A30. From the west, come off on to the A388 Western Road, heading north to Launceston. Continue into St Thomas's Road, the railway is on the left. From the east come off A30 at Lifton Down on to the A388, follow this for about 4 miles into Kensey Hill. At the cross-roads go straight on to Kensey Place. Stay on A388, after half mile turn right at cross-roads into Dockacre Road continue over another cross-roads into Wooda Road and turn right at 'T' Jct.; this is St. Thomas's Road.
Tel:01566-775665.
www.launcestonsr.co.uk
E.mail: No e.mail.
Opening times: This engine is only accessible during operating days on the railway. Check web-site for details.
Admission: Check web-site.

| Aveling & Porter | 4735 | PC 1414 | 1901 | - | TE (conv from RR) |

Leeds, Armley Mills Industrial Museum
Canal Road
Armley
Leeds
LS 12 2QF
Directions: Located 2 miles from City Centre off the A65 Kirkstall Road. Turn right into Viaduct Road which leads into Canal Road, museum is on the right.
Tel: 0113-263-7861.
www.leeds.gov.uk/museumsandgalleries
E.mail: discovery.centre@leeds.gov.uk
Opening times: Tues - Sat & Bank Hols: 10.00 - 17.00. Sun: 13.00 - 17.00.
Admission: £3.00 with concessions.

Just driving along the right road can bring results. Here an unidentified Davey Paxman portable languishes by the A456 near Kidderminster at Hodgehill Garden Nurseries on 26th August 2009.

The Museum of Lincolnshire Life has a number of steam exhibits on show and some in store; amongst the former is this early Ruston Proctor portable, No. 10743, dating from 1884 and named 'Jack'. Photographed on 16th August 2008.

There are far more Aveling & Porter rollers in museums than any other type of engine. This one was spotted at Barrow Hill engine shed on 24th August 2008. It is No. 8778 from 1916, named 'Old Faithful'.

This Merlin & Cie portable from France was spotted right alongside the A390 near St Austell. It is on private land but can easily be seen. Only one other is known on the UK mainland at the Iceni Village Museum near Swaffham in Norfolk. **Photo: Ashley Whiting.**

This rear ¾ view of a Robey tandem roller shows off the rear roll to good effect. Taken at the Great Dorset Steam Fair on 5th September 2009, the engine is normally on display at the Robey Trust in Tavistock.

The home of the North Staffordshire Traction Engine Club is at Klondyke Mill, Draycott-in-the-Clay; a number of engines are resident here and can be seen on their open days. On 13th October 2007 Thomas Green roller No. 2054 of 1927 simmers gently outside the workshops.

Preston Services near Canterbury probably houses more steam engines than any other location in the country. The list includes numerous portables and many other unusual engines as this Peerless, photographed on 26th June 2004.

Amberley Working Museum in Sussex retains a number of engines and many attend on their steam-up days. Here a Foden Wagon and a couple of Fowlers lay over during the proceedings on 17th June 2007.

58

Fowler	15428	DO 1994	1920	-	Plg
Fowler	15429	DO 1995	1920	-	Plg
Fowler (diesel)	17621	-	-	-	Plg
Shand Mason	-	-	1891	City of Leeds	FE

Leicester Museum of Technology
Abbey Pumping Station
Corporation Road
Leicester
LE4 5PX
Directions: Located to the north of the city. From M1 head for Leicester Ring Road, the A563. At Jct with A6 take 5th exit which is Abbey Lane heading for the city centre. After about ¾ mile turn left into Corporation Road, after you come off the short section of dual carriageway the museum entrance is on the right just after Wallingford Road.
Tel: 0116-299-5111.
www.leicestermuseums.ac.uk
E.mail: museum@leicester.gov.uk
Opening times: Feb - Oct: 11.00 - 16.00. Special events at other times, check web-site for details.
Admission: Free.

Note: The engine is currently in the store shed. To see it ring for an appointment.

| Aveling & Porter | 3319 | HR 6013 | 1894 | - | RR |

Leiston, Long Shop Museum
Main Street
Leiston
Suffolk
IP16 4ES
Directions: From A12 head east on B1119 to Leiston, in the town turn right at 'T' Jct then immediate left into Main Street, the museum is on the right about 200 yards along.
Tel: 01728-832189.
www.thelongshop.care4free.net
E.mail: longshop@care4free.net
Opening times: 1st Apr - 31st Oct: Mon-Sat: 10.00 - 17.00. Sun: 11.00 - 17.00. See web-site for special events.
Admission: £4.50 with concessions.

Garrett	30877	-	1912	-	Portable
Garrett	32944	BJ 3206	1916	Princess Marina	Tr
Garrett	33180	BJ 4483	1919	The Joker	Tr

Garrett	34265	SV 4945	1923	Consuelo Allen	RR
Garrett	34471	-	1925	-	Portable
Merryweather	4296	-	1901	Queen Victoria	FE

Leyland, British Commercial Vehicle Museum
King Street
Leyland
Nr Preston
Lancs
PR25 2LE
Directions: From M6 exit at Jct 28 turn west and follow brown tourist signs. The museum is approx. 1 mile along on the left near Leyland town centre.
Tel: 01772-451011.
www.bcvm.org.uk
E.mail: enquiries@bcvm.co.uk
Opening times: Apr - Sept: Sun, Tues - Thurs: 10.00 - 17.00. For other opening times see web-site.
Admission: £4.90 with concessions.

Foden	10788	AU 6695	1922	-	Wagon
Fowler	10318	WR 6790	1905	Sunny Boy 2	RL
Leyland	-	-	1923	-	Wagon
Thornycroft	No.1	-	1896	-	Wagon

Lincoln, Museum of Lincolnshire Life
Burton Road
Lincoln
LN1 3LY
Directions: Take A46 into Lincoln; follow signs 'Historic Lincoln' then 'Cathedral Quarter'. This will take you into Burton Road; the museum is on the right.
Tel: 01522-528448.
www.lincolnshire.gov.uk
E.mail: lincolnshirelife_museum@lincolnshire.gov.uk
Opening times: 10.00 - 16.00, 7 days a week. Closed Xmas & New Year.
Admission: Free.
Note: Some engines are in store but may be seen by appointment.

Fowler	16053	EB 5518	1925	John	Plg
Fowler	16054	EB 5519	1925	Michael	Plg, in store
Marshall	30169	-	1898	Harriet	Portable
Ruston, Proctor	10743	-	1884	Jack	Portable
Ruston, Proctor	46596	SFE 519L	1913	Sylvie	TE, in store
Tuxford	1131	-	1883	Maude	Portable, in store

Liverpool, Museum of Liverpool
Pier Head
Liverpool
L3 1PZ
Directions: This is the brand new Museum of Liverpool situated at Pier Head opposite the Liver building alongside the River Mersey.
Tel:0151-478-4499.
www.liverpoolmuseums.org.uk
E.mail: See web-site.
Opening times: Due to open in 2011.
Admission: To be confirmed.

Note 1: This is a new museum not yet open which will house many of the exhibits from the other museums around the city.

Note 2: Only the Sentinel wagon will be on display all other engines are to be stored for the foreseeable future.

Aveling & Porter	14025	HF 6621	1930	-	RR
Burrell	3098	MA 5657	1909	-	RL
Sentinel	6842	KA 6147	1927	-	TR
Shand Mason	-	-	1901	N. Lancs Fire Force	FP
Wallis & Steevens	2931	AA 2167	1907	-	TE

Lowestoft, East Anglian Transport Museum
Chapel Road
Carlton Colville
Lowestoft
Suffolk
NR33 8BL
Directions: Situated 3 miles SW of Lowestoft just off the A146 Lowestoft - Norwich road. Follow the brown tourist signs. From A12, also follow brown tourist signs.
Tel:01502-518459.
www.eatm.org.uk
E.mail: eastangliantransportmuseum@line.co.uk
Opening times: Various and there are special events-check web-site for details.
Admission: £6.00 with concessions.

Armstrong Whit	10R22	DX 4602	1924	-	RR
Aveling & Porter	11648	RT 2474	1926	Bor. of Lowestoft	RR
Ruston Hornsby	122282	BJ 9644	1924	Natalie 1	RR

Luton, Stockwood Discovery Centre
Mossman collection
London Road, Luton, Beds.
LU1 4LX
Directions: This site is to the SW of the town, from the M1 come off at Jct 10 and head for Luton, at Jct 10A turn left along London Road, the museum is sign-posted along here and will be found on the left after a few hundred yards.
Tel: 01582 548600.
www.stockwooddiscoverycentre.co.uk
E.mail: sdcbookings@luton.com
Opening times:
Summer: Mon - Fri 10.00 -17.00. Sat - Sun: 11 - 17.00.
Winter: Mon – Fri: 10 - 16.00. Sat - Sun: 11.00 - 16.00.
Admission: Free.

| Aveling & Porter | 6530 | NM 291 | 1908 | - | Tandem roller |
| Aveling & Porter | 10471 | NM 2570 | 1922 | Dennis | RR |

Maidstone, Kent Fire Service Museum & HQ
The Godlands
Straw Mill Hill
Tovil
Maidstone
Kent. ME15 6XB
Directions: Located within the headquarters complex, the museum is to the south-west of the town between the B2010 and the A229. From the town centre take the B2010 heading for Yalding, Straw Mill Hill is about ¾ mile along on the left. The museum is a few hundred metres along this road on the left after passing through a wooded area.
Tel: 01622-692121.
www.kent.fire-uk.org
E.mail: john.meakins@kent-fire.org.uk
Opening times: Mondays 10.00 - 16.00. Please ring prior to visiting.
Admission: Free, but donations are welcomed.

| Shand Mason | - | - | 1897 | Bromley FB | FE |

Maidstone Museum & Bentlif Art Gallery
St Faith's Street
Maidstone
Kent
ME14 1LN
Directions: Located in centre of town just east of A229. From M20 west come off at Jct 6 and take A229 Chatham Road, towards the town, continue into

Royal Engineers Road and Fairmeadow, after passing over the railway look for McKenzie Court on the left, St Faith's Street is the next on the left. The museum is on the left. From the east, take Jct 7 off M20 head south on A249 Bearsted Road, continue into Sittingbourne Road then right on to the B2012, continue on this to the Junction with the A229 and turn left into Fairmeadow, continue as above.
Tel: 01622-602838.
www.museum.maidstone.org.uk
E.mail: museuminfo@maidstone.gov.uk
Opening times: Mon - Sat: 10.00 - 17.15. Sun: 11.00 - 16.00. Closed 25/26th Dec and New Year's Day.
Admission; Free.

| Merryweather | - | - | 1902 | - | Valiant type FE |

Midleton, Co. Cork, Ireland
The Jameson Experience
The Old Distillery
Midleton
Co. Cork. Ireland
Directions: From the N25, proceed to roundabout with the R629, turn north into the town. At Junction with R626 turn left into The Rock, continue into Main Street, the distillery is along here on the right.
Tel: 353(0)21-461-3594.
www.jamesonwhiskey.com
E.mail: info@jamesonwhiskey.com
Opening times: Nov - Mar: Daily: 11.30 - 13.00 - & 14.30 - 16.00. Apr - Oct: Daily: 10.00 - 16.00.
Admission: Euros: 13.50.

| Merryweather | - | - | 1880? Midleton distillery | FE |

Milton Keynes Rural Life Museum
McDonnell Drive
Wolverton
Milton Keynes
MK12 5EL
Directions: From Jct 14 off the M1 take A500 west to first roundabout, turn right on to the A509, after 10 (yes ten!) roundabouts, having passed over a railway bridge turn right on to the A5 at the next one this is 'Abbeyhill'. Follow the brown 'cart' signs turning right into a small road continue to the next roundabout and turn right again, the museum is the first left turn.
Tel: 01908-316222.
www.mkmuseum,org.uk

E.mail: info@mkmuseum.org.uk
Opening times: Sat & Sun: Nov - to end of Mar: 11.00 - 16.30. Wed - Sun: April - to end of Oct: 11.00 - 16.30 & Bank Hols.
Admission: £5.00 with concessions. See web-site for special events.

Marshall	-	-	-	-	Portable*	
Merryweather	-	-	1912	Newport Pagnell	FE	
* Little is known about this engine; it is fitted with Clayton & Shuttleworth wheels and may originate from that manufacturer.						

Moreton-in-the-Marsh, Fire Service College

London Road,
Moreton-in-the-Marsh
Gloucestershire
GL56 0RH
Directions: Located about ¾ mile east of the town on the north side of the A44.
Tel: 01388-762191.
www.fireservicecollege.ac.uk
E.mail: See web-site
Opening times: By prior appointment only.
Admission: Free, but donations are appreciated.

Note: This is a training college and as such not generally open to the public, but by phoning groups can usually be accommodated.

| Shand Mason | 1127 | - | 1875 | - | FE |
| Shand Mason | 1909 | - | 1903 | Bor. of Huntingdon | FE |

Newby Bridge, Lakeside & Haverthwaite Railway

Haverthwaite Railway Station
Haverthwaite
Ulverston
LA12 8AL
Directions: Situated just off the A590. From the M6, turn west at Jct 36 on to the A65. At Sedgwick Turn left on to the A590 heading for Barrow, just past Newby Bridge the railway is on the right, follow brown tourist signs.
Tel: 015395-31594.
www.lakesiderailway.co.uk
E.mail: No e.mail
Opening Times: 4th April - 1st Nov.
Admission: This is an operating railway, access to the yard is free, the roller is normally on display in this area.

| Aveling & Barford | AD 073 | CUP 396 | 1938 | Westmorland Star | RR |

Newcastle Discovery Museum
Blandford Square
Newcastle-on-Tyne
NE1 4JA
Directions: Located just north of Newcastle Railway Station and immediately south of the crossroads with Westgate and Blenheim Street, off the A6082.
Tel: 0191-232-6789.
www.twmuseums.org.uk
E.mail: See web-site.
Opening times: Mon - Sat: 10.00 - 17.00. Sun: 14.00 - 17.00. Closed 25/26th Dec and New Year's Day.
Admission: Free.

| Merryweather | 3254 | - | 1887 | Prudhoe UDC | FE |

Northallerton, Prestons of Potto
Potto
Northallerton
North Yorkshire. DL6 3HX
Directions: Located in the maze of country lanes between the A19 and A172 north of where they diverge. From the A19 at a junction about 3 miles north of the divergence turn east into Trenholme Lane, after ½ mile turn right into Parsons Lane, after a left bend this becomes Parsons Back Lane. Go straight over cross-roads into Butcher Lane and into the village of Potto. At 'T' Jct turn left, the depot and museum is approx. 1 mile on the right. From the A172, about 3 miles north of the same divergence there is a cross-road, turn north-west here into Goulton Lane, follow this for about 2 miles it will bring you to the depot.
Tel: 01642-700248.
www.prestons-potto.com
E.mail: godwinjake@aol.com
Opening times: By prior appointment during Nov-June.
Admission: Free, but a donation will be appreciated.

Note: *This museum is based within a working environment so utmost care is required at the depot. Most of the engines here are operational so may well be out at rallies in the season. 10 days notice is asked for prior to any visit. There is no guarantee all or any of the engines will be on site at any one time.*

Burrell	3526	AO 6302	1913	Lightning II	Shwmns
Burrell	3618	TD 4276	1913	The Scout	Shwmns
Foster	3643	-	1908	-	Lighting machine
Ruston Hornsby	169166	TL 3433	1933	Queen Bess	TE
Savage	761	-	1900	-	Lighting machine

Norwich, Bridewell Museum
Bridewell Alley
Norwich
Norfolk. NR2 1AQ
Directions: Located just north of Norwich Castle in the city centre between London Street and St Andrews Street.
Tel: 01603-629127.
www.museums.norfolk.gov.uk
E.mail: No e.mail.
Opening times: This museum is closed for refurbishment until 2011.
Admission: To be confirmed.

| Shand Mason | - | - | 1881 | Carrow Works Fire Brigade | FE |
| Tidman | - | - | 1890 | - | OE |

Nottingham, Woollaton Park
Woollaton Park
Nottingham
NG8 2AE
Directions: Located to the west of the city between the A609 and A52. From M1 come off at Jct 25, head into the city on the A52, this is Derby Road. When you reach the interchange with the A6514 turn left on to Middleton Boulevard, head north until you reach the A609 junction, turn left on to Woollaton Road, the road divides after about 1 mile, take the left fork this is still Woollaton Road, after ½ mile turn left into Lime Tree Avenue, which leads to the Park.
Tel: 0115-915-3900.
www.nottinghamcity.gov.uk
E.mail: woollaton@ncmg.org.uk
Opening times: Apr - Oct: 11.00 - 17.00. Nov - Mar: 11.00 - 16.00.
Admission: Free.

Croskill	UP5	-	-	-	Portable
Fowler	17756	VO 8988	1929	-	Plg
Fowler	17757	VO 8987	1929	-	Plg
J.T. Marshall	-	-	1886	-	Portable

Oxford City Fire & Rescue Service
Fire Station
Rewley Road
Oxford
OX1 2EH
Directions: From the A34 southern by-pass come off at the interchange with the A420 and turn east for Oxford, at the 'T' Jct turn left into Botley Road and continue along here going under a railway bridge. Take the 2nd left after the

bridge into Rewley Road; the fire station is along here on the right.
Tel: 01865-242323.
www.oxfordshire.gov.uk
E.mail: petergraham@oxfordshire.gov.uk
Opening times: This is an operational fire station not a museum, so please ring the above number before visiting. Ask for the Officer in Charge.
Admssion: Free.

Note: *This engine may be moving to Greenock to a new museum with the Strathclyde Fire Preservation Group. See listing elsewhere.*

Merryweather	-	-	1900? Victoria	FE

Peastonbank, Glenkinchie Visitor Centre
Glenkinchie Distillery
Pencaitland
East Lothian
EH34 5ET
Directions: Situated 15 miles south west of Edinburgh between the A1 and A68 trunk roads, due south of Pencaitland in the village of Peastonbank.
Tel: 01875-342004.
www.discoveringdistilleries.com
E.mail: Rhona.paisley@diageo.com
Opening times: Summer: Mon - Sat: 10.00 - 16.00. Sun: 12.00 - 16.00.
Admission: £5.00 with concessions.

Merryweather	-	-	1940? -	FP

Poole Museum
4 High Street
Poole
BH15 1BW
Directions: Enter Poole via the A350, West Street until you reach the Sterte Roundabout, continue on the A350 for another ¼ mile then turn left into New Orchard. Take the third right into High Street, continue to the end where it bears left. The museum is in front of you.
Tel: 01202-262600.
www.poole.gov.uk
E.mail: museums@poole.gov.uk
Opening times: 21st Mar - 2nd Nov: Mon - Sat: 10.00 - 17.00. Sun: 12.00 - 17.00. 3rd Nov - 5th Apr: Tues - Sat: 10.00 - 16.00. Sun: 12.00 - 16.00.
Admission: Free.

Shand Mason	-	-	1892 Victor	FE

Portsmouth City Museum
Museum Road,
Portsmouth
Hants.
PO1 2LJ
Directions: The museum is right in the city, from the end of the M275 continue on to the old A3 which is Mile End Road, go straight on at the first roundabout and turn right at the next and straight on again at the third, this is Alfred Road which leads into Anglesea Road, follow the one way system staying on the A3 into St Michaels Road. At the crossroads with the B2154 turn right into Museum Road, the museum is on the left, parking is free on site.
Tel: 023-9282-7261.
www.portsmouth.gov.uk
E.mail: mus@portsmouthcc.gov.uk
Opening times: Apr - Sept: 10.00 - 17.30. Oct - Mar: 10.00 - 17.00. Closed 24 - 26th December.
Admission: Free.

| Aveling & Porter | 10312 | BK 6748 | 1922 | - | RR |

Note: This engine may now be moved to a new location

Preston, Lancashire Fire & Rescue Service

Note: The Shand Mason stored here is not on display and is likely to be moved to the British Commercial Vehicle Museum at Leyland at some future point

| Shand Mason | - | - | 1880 | Fulwood UDC | FE |

Ramsey Rural Museum, Cambs
Wood Lane
Ramsey
Cambs. PE26 2XD
Directions: Located just to the north end of the village to the east of the B1096. Turn right opposite a church on your left into Wood Lane.
Tel: 01487-814304.
www.ramseyruralmuseum.co.uk
E.mail: d.yardley@talk21.com
Opening Times: Apr - Oct: 10.00 - 17.00. Sun & Bank Hols: 14.00 - 17.00.
Admission: £4.00 with concessions.

| Savage | 561 | - | - | 1892 | - | CE |

Reading, Englefield Estate
Englefield Road

Theale
Nr Reading
RG7 5DU
Directions: From M4 Jct 12 take A4 west to Newbury, go straight over first roundabout, at next take A340 north to Pangbourne, after ½ mile, turn left into Englefield Village, follow signs to the Estate.
Tel: 0118-930-2504.
www.englefieldestate.co.uk
E.mail: office@englefield.co.uk
Opening times: See web-site for details.
Admission: See web-site for details. Please phone to check accessibility at this site.

Shand Mason	-	-	1894	Englefield	FE

Reading, Museum of English Rural Life
22 Redlands Road
Reading
RG1 5EX
Directions: Located near the centre of Reading. From the M4 come off at Jct 10 and go north on the A329 heading for the town. At a roundabout with the A4 turn left into London Road, after approx 1 mile turn left into Redlands Road, the museum is on the right along here.
Tel: 0118-378-8660.
www.reading.ac.uk/merl
E.mail: merl@reading.ac.uk
Opening Times: Tues - Fri: 09.00 - 17.00. Sat & Sun: 14.00 - 16.30. Closed Bank Hols, Christmas, Easter and University closure dates.
Admission: Free, but donations are welcome.

Clayton & Shutt.	15635	-	1877	-	Portable

Rochdale, Greater Manchester Fire Service Museum
Maclure road
Rochdale
OL11 1DN
Directions: From M62 come off at Jct 20 and head north on A627(M) into Rochdale. At Jct with A664 Edinburgh Way, swing left along to the next roundabout and then turn right into A58 Manchester Road. At traffic lights turn right into Tweedale Street and go along there to roundabout where you need to turn left into Maclure Road, the museum is along here on the right.
Tel: 01706-901227.
www.manchestfire/gov.uk/museum
E.mail: museum@manchesterfire.gov.uk

Opening times: 10.00 - 16.00 every Friday and first Sunday of each month
Admission: Free.

| Shand Mason | - | - | 1910 | George V | FE |

St Austell
This Portable is located alongside the A390 east of St Austell on the way to Lostwithiel. It is on private land but can be seen from the roadside.

| Merlin & Cie | - | - | - | - | - | Portable |

Saltash, St. Dominick, Cotehele House,
St. Dominick
Nr Saltash
Cornwall
PL12 6TA
Directions: Located in the maze of country lanes between the A388 and A390, 1½ miles east of St Dominick alongside the River Tamar.
Tel: 01579-351346.
www.nationaltrust.org.uk/main/w-cotehele
E.mail: cotehele@nationaltrust.org.uk
Opening times: These vary consult web-site for more information.
Admission: £9.50 with concessions. Garden & Mill only £5.50.
National Trust members Free.

| Merryweather | 2142 | - | 1903 | - | Fire Pump |

Scarborough, The Scarborough Fair Collection
Flower of May Holiday Park
Lebberston Cliff
Scarborough
North Yorks
YO11 3NU
Directions: Situated 2 miles north of Filey and 5 miles south of Scarborough just off the A165. At the roundabout with the B1261 (to Cayton) turn east after passing the Plough pub on your left, at the next 'T' Junction turn right, The holiday park is about ½ mile along here on the left.
Tel: 01723-586698.
www.scarboroughfaircollection.com
E.mail: info@scarboroughfaircollection.com
Opening times: Sundays: 12.00 - 17.00 Mar - Nov.
Admission; £5.00 with concessions.
Note: *All the engines in this collection are operational and thus can be out at rallies, particularly in the summer months.*

Burrell	3444	CK 3403	1913 His Lordship	Shwmn
Burrell	3878	BT 3997	1921 Island Chief	Shwmn
Fowler	15657	FX 6661	1920 The Iron Maiden	Shwmn
Garrett	33284	DX 3099	1918 Princess Maud	Sh, Tr
Savage	607	-	1894 Granny	CE

Scunthorpe, Normanby Hall Country Park
Normanby by Scunthorpe
North Lincolnshire
DN15 9HU
Directions: Located 4 miles north of Scunthorpe just off the B1430, follow brown tourist signs. It is also sign-posted off the Humber Bridge and the A181.
Tel:01724-720588.
www.normanbyhall.gov.uk
E.mail: normanbyhall@northlincs.gov.uk
Opening times: The engine is in the Coach House which is open from April - September. The house itself is open all year round.
Admission: £5.00 with concessions.

| Shand Mason | - | - | 1900 | Normanby House | FE |

Sheffield Fire & Police Museum
Old Fire Station
West Bar
Sheffield
S3 8PT
Directions: Located at the intersection of the A61 and A57 in the city at West Bar. The museum is right on the corner of the roundabout.
Tel: 0114-2491-999.
www.firepolicemuseum.gov.uk
Email: info@firepolicemuseum.org.uk
Opening times: Sun & bank Hols 11.00 - 17.00.
Admission: £4 with concessions.

| Shand Mason | - | - | 1882 | Horrocks's Norwich | FE |
| Merryweather | - | - | - | Trench steamer | Valiant Type |

Sheffield, Kelham Island Museum
Alma Street
Off Corporation Street
Sheffield
S3 8RY
Directions: Come off the M1 on to the A57 Parkway to Park Square roundabout. Turn right into Exchange on the one-way system, then left into Castle Gate which continues into Bridge Street, at the next roundabout go

straight on into West Bar. At the next, turn right into Corporation Street then third left into Alma Street, the museum is off here on the right.
Tel: 0114-272-2106.
www.simt.co.uk
E.mail: postmaster@simt.co.uk
Opening times: Mon - Thurs: 10.00 - 16.00 all year. Closed Fri & Sat.
Admission: Free in school holidays at other times £4.00 with concessions.

Fowler	7232	WR 6662	1894	-	TE

Sheffield, Renishaw Hall
Ecklington
Derbys
S21 3WB
Directions: From Jct 30 off the M1 head west on A6135 Sheffield Road, this continues as Main Road. Renishaw Hall is on the left along this road.
Tel: 01246-432310.
www.renishaw-hall.co.uk
E.mail: info@renishaw-hall.co.uk
Opening times: 2nd Apr - 27th Sept: 10.30 - 16.30: Thurs, Fri, Sat, Sun & Bank hols.
Admission: £5 with concessions (car park £1).

Ruston, Hornsby	-	-	1921?	-	Portable

Sittingbourne & Kemsley Light Railway
Sittingbourne
Kent
ME10 2DZ
Directions: From M2 or M20 take Jct 7 follow A249 then go on to A2 to Sittingbourne town centre and railway station. The S & KLR station is close by the Sittingbourne retail park; brown tourist signs will direct you.
Tel: 0871-222-1568.
www.sklr.net
E.mail: info@sklr.net
Opening Times: See above, check web-site for details.
Admission: Check web-site for fares and access to Kemsley station & yard

Aveling & Porter	10784	PU 3094	1924	Suzie	RR
Aveling & Porter	-	-	-	-	RR

Note: *The engines are at Kemsley yard at the far end of the line. It is possible to drive to within about ½ mile of it, thence proceed by foot. The other way is to take the train from Sittingbourne. However the line is temporarily closed due to commercial difficulties but is expected to re-open in 2010.*

Sittingbourne, Bredgar & Wormshill Railway
The Warren
Bredgar
Sittingbourne
Kent. ME9 8AT

Directions: From M20 take Jct 8 on to A20 towards Ashford, at 2nd roundabout turn left on to a small road to Hollingbourne village, after the village there is a steep hill; carry on for about 4 miles to the railway. From M2 take Jct 5 head for Maidstone on A249, after a short way turn left for Oad Street, go through the village, at the 'T' Jct turn right for Bredgar, go on until you see the village pond then turn right, the railway is about 1 mile along on the left.
Tel: 01622-884254.
www.bwlr.co.uk
E.mail: willambest@btinternet.com
Opening times: 1st Sunday of the month July - October: 10.00 - 17.00
Admission: £7.50.

Note: This is a preserved railway with a number of traction engines on site.

Burrell	2551	-	1903	-	RR
Garrett	33305	HT 7112	1918	Mighty Atom	TR
Garrett	33442	BL 9009	1919	Caroline	TE
Ruston & Hornsby	115023	XM 6374	1922	Veronica	RR

Skegness, Church Farm Museum
Church Road South,
Skegness
Lincs
PE25 2HF

Directions: Situated between the A158 and the A52, to the west of the town centre. From the A158 Wainfleet Road about ¼ mile after coming into the built-up area take 1st right into Lincoln Road, this gradually swings left, then straightens out, after ¼ mile turn right into Queens Road, then 1st left into Church Road South, the museum is on the right. From the A52 after passing a caravan site on your left, take 1st left into Queens Road continue for ½ mile to Church Road South which is on your right.
Tel: 01754-766658.
www.skegness-resort.co.uk
E.mail: churchfarmmuseum@lincolnshire.gov.uk
Opening times: 10.00 - 17.00 during summer season.
Admission: Free, (except for special events) see web-site,

| Richard Hornsby | 7297 | - | 1892 | Bob | TE |

Southampton, Bursledon Brickworks
Coal Park Lane
Swanwick
Southampton
SO31 7GW

Directions: Located adjacent to the M27 east of Southampton in village of Swanwick. Come off M27 at Jct 8 turn south into Bursledon Road, at next roundabout turn left on to A27 Providence Hill, heading for Fareham. Stay on the A27 go over the creek and railway bridges and take first right into Swanwick Lane, after approx 500mtrs turn left into Coal Park Lane.
Tel: 01489-576248.
www.fareham.gov.uk
E.mail: enquiries@hampshirebuildings.org.uk
Opening times: Sundays 10.00 - 18.00. Thursdays: 13.00 - 16.00. There are also special events most months. See web-site for details.
Admission: £3 but may vary on special events.

Note: *This is a private collection of engines which may change from time to time. They are only accessible during official open days, when additional engines are often present. See web-site for details.*

Davey Paxman	-	-	-	-	Portable
Garrett	29764	AC 9326	1911	Olive	TE
Marshall	84480	-	1929	-	Portable

Southwark, London Fire Brigade Museum
Southwark Fire Station
94A Southwark Bridge Road
London
SE1 0EG

Directions: Just south of the Thames, the nearest Tube station is 'Borough' on the Northern line. From there walk south along Borough High Street after 200 mtrs turn right into Lant Street continue to the cross-roads with Southwark Bridge Road then turn right, the museum is on the left along here. By bus take the 344.
Tel: 020-8555-1200 ext 39894.
www.london-fire.gov.uk
E.mail: museum@london-fire.gov.uk
Opening times: Visits by appointment only Mon - Fri: 10.00 or 14.30.
Admission: £3.00.

Shand Mason	No.48	-	1891	Victoria	FP, not on show
Shand Mason	-	-	1891	Rickmansworth	FE
Shand Mason	-	-	-	P3 (known as)	FE

South Kensington Science Museum
South Kensington
London
SW7 2DD
Directions: Parking is difficult so use the Tube to South Kensington Station or bus 14, 49, 70, 74, 345, 360, 414, or C1 and follow signs to museum.
Tel: 0870-870-4868.
www.sciencemuseum.org.uk
E.mail: foi@nmsi.ac.uk
Opening Times: 10.00 - 18.00 every day except 24 - 26th December.
Admission: Free.

| Aveling & Porter | 721 | KL 3631 | 1873 - | TE |

Stoke, Potteries Museum & Art Gallery
Bethesda Street
Hanley
Stoke-on-Trent
ST1 3DW
Directions: From M6 take Jct 15 or 16 on to the A500 to Stoke, Follow signs for City Centre (Hanley) Cultural and Potteries Museum. Parking is 5 minutes away. There is a drop-off point in Bethesda Street.
Tel: 01782-232323.
www.stoke.gov.uk
E.mail: museums@stoke.gov.uk
Opening times: Mar - Oct: Mon - Sat: 10.00 - 17.00. Sun. 14.00 -17.00. Nov - Feb: Mon - Sat: 10.00 -16.00. Sun: 13.00 - 16.00.
Admission: Free.

| Merryweather | 1305 | - | 1894 - | FE |

Stowmarket, Museum of East Anglian Life
Crowe Street
Stowmarket
Suffolk
1P14 1DL
Directions: From A14 going north come off at junction with A1120 turn left on to that, heading for Stowmarket, at 2nd roundabout turn right into Needham Road. At next roundabout continue on straight into Gripping Way, at cross-roads with Station Road West turn left, this is the B1115, at next cross-roads go straight on into Tavern Street, which leads into Finborough Road. After ¼ mile turn left into Liffe Way, 200 yards on turn left into Asda car park. Entrance to the museum is in the far corner of the car park.
Tel: 01449-612229.

www.eastanglianlife.org.uk
E.mail; via web-site
Opening times: 22nd Mar - 31st Oct: Mon - Sat: 10.00 - 17.00. Sun: 11.00 - 17.00.
Admission: £6.50 with concessions. In winter, £2.

Burrell	776	-	1879	The Countess	Plg
Burrell	777	-	1879	The Earl	Plg
Burrell	3399	AH 5446	1912	The Empress of....	TE
RSJ	5137	Z 1	1882	Lizzie McKenna	TE
Turner E. R & F	981	-	1874	Mary	Portable
Walsh & Clarke	5038	-	1919	-	Plg*
Walsh & Clarke	5039	-	1919	-	Plg*

*Note: these are actually paraffin powered but were built to look like steam engines, the boiler being used as the fuel tank.

Stradbally Steam Museum
The Green
Stradbally
Laois
Eire
Directions: Located on the main N80 route through the town, in the town itself.
Tel: 057-862-5114.
www.stradballysteammuseum.org
E.mail: fenlonsean@yahoo.co.uk
Opening times: Mon - Fri: 09.30 - 17.30. Sundays and Bank holidays: 14.00 - 17.00 by prior appointment.
Admission: Check on arrival.

Aveling & Porter	8563	IY 1682	1915	-	RR
Fowler	21649	KI 1973	1936	-	TE
Mann	1216	U 4249	1917	Mighty Mann	Tr

Strumpshaw Steam Museum
Strumpshaw Old Hall
Strumpshaw
Norwich
NR13 4HR
Directions: Located to the south of the A47 Norwich - Yarmouth main road. At the Brundall roundabout take 1st left (alongside a petrol station) continue to traffic lights and turn right into Stocks Lane, carry on to a mini roundabout and turn left into Strumpshaw Road and on into the village. Turn right into Stone Road, after ½ mile turn right into Low Road follow signs to museum.
Tel: 01603-714535.
www.strumpshawsteammuseum.co.uk
E.mail: strumpshawmuseum@aol.com

Opening times: Variable so please check web-site.
Admission: £4.00.

Aveling & Porter	6221	EB 02	1907 -		TR
Aveling & Porter	10079	BJ 7116	1921 -		RR
Aveling & Porter	12461	MY 953	1929	Princess Anne	TR
Aveling & Porter	14085	VF 9891	1930 -		RR
Aveling & Porter	14135	NG 1412	1931 -		RR
Burrell	2366	AH 5488	1901	Buller	TE
Burrell	3789	SW 742	1918	Princess Royal	TE
Burrell	3924	NO 6085	1922	Princess Mary	TE
Foden	13802	VN 2912	1931	Santa Maria	Wagon
Fowler	15340	BL 8167	1919 -		Plg
Fowler	15341	BL 8168	1919 -		Plg
Garrett	34045	BJ 6628	1921	Lou Lindy	TE
Marshall	28334	EB 6519	1897	Laura	TE
Marshall	52483	E 5114	1909 -		TE
Marshall	66182	CT 3926	1914	Prince Andrew	TE
Marshall	83270	VH 1261	1929	Jolly Joyce	RR
Marshall	83600	VF 3665	1928 -		TE
Marshall	83780	VF 4183	1928 -		TE
RSJ	36030	-	1925 -		Portable
Ruston Proctor	38994	AH 7938	1910	The Champion	TE
Wallis & Steevens	7801	PU 3378	1924	Wanda	TE

Swaffham, Iceni Village Museum

Cockley Cley
Swaffham
Norfolk
PE37 8AG
Directions: Situated a couple of miles south of Swaffham west of the A1065 Brandon road. Turn off the main road to Cockley Cley following the brown tourist signs.
Tel: 01760-724588.
www.icenivillage.com
E.mail: No e.mail
Opening times: April - October: 11.00 - 17.00. (17.30 in July & August).
Admission: Check web-site

Merlin & Cie	-	-	-	-	Portable

Swansea, South Wales Fire & Rescue Service

Merryweather	2662	-	1907? -	FE

Note: This engine is currently not on display

Swansea Waterfront Museum
Oystermouth Road
Maritime Quarter
Swansea
SA1 3RD
Directions: From M4 west-bound exit Jct 42 or from the east Jct 47. Head south following the brown tourist signs into Swansea. The Museum is right by the waterfront. There is a large car-park close by.
Tel: 01792-638950.
www.museumswales.ac.uk
E.mail: info@museumswales.ac.uk
Opening times: Daily 10.00 - 17.00.
Admission: Free.

Aveling & Porter	8370	NK 4562	1914	-	RR

Swansea, Welsh Museum of Fire
Unit 31
Lonlas Village Workshops
Skewen
Neath
SA10 6RP
Directions: Located north of Swansea adjacent to the M4. From the east come off M4 at Jct 43, turn left on roundabout and then on to B4290, immediately turn right into Pen-Yr-Heol, staying on the B4290 into Burrows Road, At 'T' Jct with A4230 turn left almost doubling back on yourself. After ½ mile Lonlas Road is on the right, the museum is just along here on the left. From the west, exit M4 at Jct 44 turn right into B4291, at 'T' Jct turn right again staying on B4291, continue onto A4230, Lonlas Road is about 1 mile further on the left.
Tel: 01639-635761.
www.wafersmuseum.org.uk
E.mail: devans@btinternet.com
Opening times: Prior arrangements must be made to visit this museum.
Admission: Free, but donations are appreciated.

Shand Mason	-	-	1912 Tenbury	FE

Tamworth, Statfold Barn Railway
Ashby Road
Tamworth
Staffordshire
B79 0BU

Directions: Located just north-west of Tamworth in a 'V' between the M6 toll road and M42 on the B5493 halfway between Tamworth and Seckington. Come off M42 at Jct.11 turn west on to B5493, pass through Seckinton, the railway is about 1 mile along here on the left. From M6 toll exit Jct 3 on to A38 northbound, at the roundabout turn right on to A453 to Tamworth. Turn right on to A5 and exit at first Jct., head north straight over 2 roundabouts on to the A51. At next junction turn right on to Lichfield Street which is the B5493, turn left on to Aldergate still on the B5493, go over railway into Upper Gungate (A513) turn right on to B5493 again, Ashby Road, the railway is 1½ miles on right.
Tel: No phone number.
www.statfoldbarnrailway.co.uk
E.mail: henry@statfold-oils.co.uk
Opening times: This is a private site and has limited opening times; those planned for 2010 are 27th March, 5th June and 18th September. Check web-site for an invitation form which must be completed prior to any visit.
Admission: A contribution of £8.00 is appreciated.

Fowler	13482	NO 797	1913	-	Plg
Marshall	84679	WF 2657	1929	Brother Bob	TE
Wallis & Steevens	7939	OT 6514	1927	Emily	RR Simp.

Tavistock, The Robey Trust

The New Perseverance Iron Works
Crelake Industrial Estate
Pixon Lane
Tavistock
Devon
PL19 9AZ

Directions: Situated in Tavistock itself. From the A386 Plymouth Road, if coming north a landmark is a Texaco garage 200yds past which turn right into Pixon Lane, under the old railway bridge and then turn right. If coming south, go through the town, note the cattle market off the Whitchurch Road and turn right into Pixon Lane (the other end) immediately in front of the 'Market Inn' then left before the railway bridge. From the Callington direction, join A386 for Plymouth and turn left into Pixon Lane before Texaco garage. Follow the sign for 'King's', after the 'S' bends the works is on the right, adjacent to the King's Community Centre.
Tel: 01822-615960 Dave Davies.
Or: 01822-854108 Norman Emmett.
www.therobeytrust.co.uk
E.mail: No e.mail
Opening times: By appointment only, phone either of the above numbers.
Admission: Free, but a donation would be welcome.

Note: This is not a museum but a workshop, so for Health & Safety reasons please take care, particularly with children. You will be made very welcome but please respect the above statement.

Robey	31821	-	1913	-	Tr
Robey	32387	DE 2592	1913	-	RR
Robey	33348	Z 200	1914	-	TE
Robey	40991	FE 5373	1923	Shamrock	Conv.
Robey	42693	FE 7490	1925	Stumbles	Tandem RR
Robey	45655	VL 2773	1930	Herts Wanderer	Tri-Tandem RR
Robey	57193	-	1955	-	Steam producer
Robey	-	-	1875?	-	Portable

Thetford, Burrell Museum
Minstergate
Thetford
Norfolk
IP24 1BN

Directions: From A11 Thetford by-pass eastbound, turn right at roundabout into the A134 Brandon Road. Take the 5th left into Bridge Street go over river and turn left into Minstergate. From A11 westbound, come off at roundabout with A1075 head into Thetford, at next roundabout go straight on into Norwich Road, take 4th left into Earls Street at 'T' Jct turn right and at roundabout immediately ahead of you turn right into Minstergate. The museum is on the left with a car park just past it.
Tel: 01842-765840.
www.culture24.org.uk
E.mail: burrell@thetfordtowncouncil.gov.uk
Opening times: Tues: Apr - Oct: 10.00 - 14.00.
Last Sat of month Mar - Nov 10.00 - 16.00.
Admission: Free.

Burrell	748	CF 3667	1877	Century	TE
Burrell	2479	AH 5239	1902	-	TE*
Burrell	2626	TB 3717	1903	Spitfire	TE*
Burrell	2948	NO 106	1907	Dreadnought	TE*
Burrell	3106	NO 439	1909	Princess Royal	TE*
Burrell	3622	NO 306	1914	General French	TE*
Burrell	3695	TB 2740	1915	Lord Derby	TE
Burrell	3833	BR 1498	1920	Queen Mary	Shwmn*
Burrell	4061	UO 945	1927	-	RR*

Note: engines marked * are often out on loan and can be seen at local rallies.

Thirsk, Ampleforth Farming Flashback
Thorpe Hall Farm
Ampleforth
Yorkshire

YO62 4DL
Directions: Located east of Thirsk off the A19 and south of the A170. Turn east off the A19 at Easingwold along some minor lanes heading for Crayke. At Junction in the village turn left for Brandsby. Once in the village turn left for Yearsley along Yearsley Moor Bank. Museum is about 1 mile on left. From the A170 turn south on to B1257 at Sproxton continue to Oswaldkirk, turn right on to B1363 through Gilling East, after 3 miles turn right to Yearsley, turn right at cross-roads in village. Museum is about 1 mile on left.
Tel: 01439-788793.
www.yorkshire.com
E.mail: No e.mail
Opening times: Apr - Sept: Mon - Sat: 10.00 - 18.00.
Admission: £2.50.

| Burrell | 4032 | PW 6287 | 1925 | Firefly | TE |

Watford Fire Station & Museum
Whippendell Road
Watford
Herts
WD18 7QW
Directions: From M1 take Jct 5 and go west on the A4008 heading for Watford along Stephenson Way. At first roundabout go straight on bear left at next junction into Exchange Road follow road around a long right hand bend, turn left at 'T' Jct into Beechan Grove, then right into Cassio Road. At 'T' Jct turn left into Rickmansworth Road after about ¾ mile the fire station is on the left just prior to a roundabout.
Tel: 01923-232297.
www.watfordmuseum.org
E.mail: firemuseum@watfordmuseum.org.uk
Opening times: Wednesday 9.30 - 11.30. There are occasional special events on Saturdays through the summer months, so check web-site for details.
Admission: Free.

| Shand Mason | - | - | 1896 | Hertfordshire Fire Brigade | FE |

Note: *This museum will be moving in the spring of 2010 to a new site opposite the Watford Town Museum in the High Street to where the old Sedgwicks Brewery was once located.*

Weedon, National Fire Service Museum
Cavalry House
Building 86
The Old Depot

Bridge Street
Weedon
Northants
NN7 4PS

Directions: The museum is located in the old Ordnance Depot that straddles the Grand Union Canal. Head for the intersection of the A5 and A45, the turning to the museum is about 200mtrs south of this junction on the A5 turn west into High Street, progress into Bridge Street which goes under the canal and the railway, at the next junction turn right into Harmans Way, the museum entrance is on the left about 100 mtrs along here.
Tel: 01327-342396.
www.friends-of-fireworld.com
E.mail: patmurfin@msn.com
Opening times: This museum is still in the throes of construction and many exhibits are not normally on public view. If you are wishing to visit please contact the above first to ascertain the current situation.
Admission: To be confirmed.

Merryweather	-	-	1885 Bexhill	FE

Wells-next-the-Sea, Holkham Hall

Holkham Hall Bygones Museum
Wells-next-the-Sea
Norfolk
NR23 1AB

Directions: 2 miles west of Wells on the A149, follow brown tourist signs.
Tel: 01328-710227.
www.holkham.co.uk
E.mail: enquiries@holkham.co.uk
Opening times: Easter - October; daily 10.00 - 17.00.
 Note: opening times for the Hall itself vary, please check web-site.
Admission: £5.00.

Farmers Foundry	39	-	1910 -	Portable
Tidman	-	-	1870? -	CE

Westonzoyland Pumping Station

Hoopers Lane
Westonzoyland
Nr Bridgwater
Somerset. TA7 0LS

Directions: Located by the River Parrot, 2 miles from Westonzoyland and 3 miles from Burrowbridge. Leave M5 at Jct 23 head into Bridgwater and turn left

The Burrell Museum at Thetford exhibits a variety of that company's engines, as seen here from the mezzanine gallery. No.2479 stands in front of No.748 with 3833 'Queen Mary' at the rear on 28th April 2007.

This Ruston Hornsby roller normally resides at the East Anglian Transport Museum at Carlton Coleville near Lowestoft. She was built in 1924 as Works No. 122282, Reg: BJ 9644.

The National Railway Museum at York houses a couple of Merryweather fire engines, this photo taken on 23rd May 2009 is of 'Gateshead' dating from the 1880s. The other is a 'Valiant' type.

Shand Mason fire engines are well represented around the country, This one, 'Thorney' can usually be found at the Bedfordshire Fire Station at Biggleswade. On the 11th October 2009 she was fully rigged out at the nearby Stotfold Mill.

Thought for many years to be a Shand Mason it is now certain that this is actually a rare William Rose engine dating from around 1899. Seen here at Burton-on-Trent fire station on 1st December 2009.

The Scarborough Fair collection is unusual in that virtually all the engines are steamable and can frequently be seen out at rallies. At Pickering on 8th August 2009, Burrell Showman 3878 of 1921, Reg: BT 3997, named 'Island Chief' certainly appears to have has her fair share of coal to steam up on!

Hollycombe House has a large collection of engines, here, just two of them, basking in the sun are W & S No. 8023 'Simplicity' roller, Reg: CG 571 from 1932, converted to burn petrol or paraffin and Mann tractor No 1260 of 1917, Reg: U 4298 on 1st July 2006.

An unlikely location to find Traction engines is Kemsley station yard at the Sittingbourne & Kemsley Light Railway. These two rollers were spotted there photographed on 28th September 2008, the nearest being Aveling & Porter 14044 of 1930.

86

on to A372 to Lamport. Follow signs into Westonzoyland; note a small shop on the right and then turn into School Lane, there should be a sign pointing to the pumping station. Road then bears hard left but you need to go straight on into narrow lane called Lakewall, after 1 mile bear left at fork into Hoopers Lane, museum is on right along here.
Tel: 01275-472385.
www.wzlet.org
E.mail: steamteam@wzlet.org
Opening times: Suns and Bank hols: 13.00 - 17.00.
Admission: £3.00 on non-steam days, £4.00 on steaming days.

Marshall	88553	-	1938	-	Portable

Weybourne, North Norfolk Railway
Station Approach
Sheringham
Norfolk
NR 26
Directions: The engine is located in Weybourne station yard just north of the A148 between Holt and Sheringham and can just be seen from the road. From the west go through Holt and High Kelling after mile or so turn left into Gipsies Lane, the railway is signposted from here. From the east turn right off the A148 at the same point. Follow this lane until you go over a railway bridge. The Station and yard are immediately after on your right.
Tel: 01263-820800.
www.nnrailway.co.uk
E.mail: enquiries@nnrailway.co.uk
Opening times: See web-site for operating times.
Admission: Free to view engine in yard.

Foster	14422	CT 6708	1924	-	TE

Note: this engine is privately owned but can be seen from the road or the station yard

Winchester, Chilcomb House
Chilcomb Lane
Winchester
SO23 8RD
Directions: From M3 northbound take Jct 10 sign-posted; Winchester Bar End, at roundabout with traffic lights take 2nd exit into Bar End Road (A3404). Pass over a bridge and turn almost immediate right into Chilcomb Lane. From M3 southbound come off at Jct 9 on to A272 (A31) to Winchester, stay on A272 which runs parallel to M3, at the roundabout go on to A31, at next roundabout go left, still on A31, pass under M3, at roundabout with traffic lights take 2nd exit into Bar End Road go over bridge and right into Chilcomb Lane.

Tel: 01962-826700.
www.hants.gov.uk/museums/
E.mail: See web-site
Opening Times: Chilcomb House is a store for the Hampshire Museums Service and as such is not normally open to the public. However it may be possible to obtain access by making an appointment.
Admission: Free.
Note: Engines are occasionally exchanged between here and Milestones Museum in Basingstoke.

Shand Mason	-	-	- Lymington	FE
Tasker	203D	-	1878 -	Stationary Boiler
Tasker	352	NO 1060	1893 Excelsior	TE
Tasker	1643	AA 5506	1915 -	Motion and top works only
Tasker	1675	AP 9027	1915 -	TE
Tasker	1715	BP 6289	1916 -	RR Class A2
Tasker	1776	AP 9281	1918 -	TE Class C
Tasker	1818	HO 2822	1918 -	Cut away parts only
Wallis & Steevens	7614	AF 3456	1919 - Sir Douglas	In restoration

Wroughton, Science Museum
Redbarn Gate
Wroughton
Swindon
Wiltshire. SN4 9LT
Directions: From the east come off the M4 at Jct 15, head south on the A346, turn right on to the B4005 for Wroughton. At 'T' Jct with A4361 turn left, museum is about 1 mile further on the left. From the west come off the M4 at Jct 16 on to A3102 immediately turn left on to B4005, Hay Lane. At 'T' Jct in Wroughton turn right and continue on the A4361 as above.
Tel: 01793-846226 for individual research requests.
Tel: 01793-846205 for booking group tours of Hangar D4 only on 1st Wednesday of months: May - Oct.
www.sciencemuseum.org.uk
E.mail: wroughtonenquiries@sciencemuseum.org.uk
Opening times: Only open on special occasions (see above) and for individuals and pre-booked parties Mon-Fri only 10.00 - 16.00.
Admission; Free for special occasions and individual researcher access.

Note 1: This is a store for the Science Museum's other museums around the country and as such not normally open to the public except under the above provisions.
Note 2: Special provision **may** be granted to allow access for groups to view the steam engines in storage but a fee will be charged and it will be subject to availability of staff. Bookings should be made 4-6 weeks in advance. There are no refreshments facilities on site.

							Mus. Ref No.
Aveling & Porter	2185	BY 2185	1886	-		RR	1966-18
Foden	1174	SN 1609	1906	Pride of Leven		Wagon	1982-336
Foden	6368	M 8118	1916	Pride of Edwin		Wagon	1982-337
Fowler	15194	FX 6820	1918	William Shakespeare		Plg	1984-2226
Fowler	15195	FX 6821	1918	Jane Hathaway		Plg	1984-2227
McLaren	112	KL 2176	1881	Empress of India		TE	1985-952
Merryweather	-	-	1863	Sutherland		FE	1924-211
Merryweather	-	-	1902	LCC Fire Brigade		FE	1920-21
Richard Hornsby	1851	-	1871?	-		Portable	1968-606
Sentinel	1718	AW 3835	1917	-		Wagon	1985-95
Shand Mason	-	-	1894	Southgate DC		FE	1945-3
Tidman	-	-	1900	-		OE*	?
Tidman	-	-	1900	-		CE*	?
Walsh & Clark	-	-	1919?	-		Plg**	1982-1035

Note: *The whereabouts of these two engines is still unconfirmed.*
****Note:*** *This is actually paraffin powered but has been built to look like a steam traction engine, the boiler being used as the fuel tank.*

York Castle Museum
Tower Street
York
YO1 9RY
Directions: This is in the centre of York just east of the river between Bridge Street and Bishopsgate bridges on the corner of Tower Street and Clifford Street and adjacent to Coppergate Shopping Centre. There is 3 hours parking at the side of the museum.
Tel: 01904-653611.
www.yorkcastlemuseum.org.uk
E.mail: enquiries@ymt.org.uk
Opening times: 09.30 - 17.00: daily. Closed 25/26th December and New Year's Day.
Admission: £7.50 with concessions.

Shand Mason	-	-	1904	Arkle	FE

York, National Railway Museum
Leeman Road
York
YO26 4XJ
Directions: Head into York itself on the A1036 from the north or south. As you near the railway station follow the brown tourist signs to the museum, there is parking on site. By train, there is a footpath bridge direct from the station into the museum.
Tel: 08448-153139.
www.nrm.org.uk
E.mail: nrm@nrm.org.uk

Opening times: 10.00 - 18.00 daily. Closed 24 - 26th Dec.
Admission: Free.

| Merryweather | - | - | 1880? Gateshead | FE |
| Merryweather | 1952 | - | 1901 GWR 165 | Valiant type |

Note: the engines are in the Station Hall on the south side of Leeman Road.

Wallis & Steevens Wagon No. 7279, Reg: AA 2470 built in 1912, was photographed at Milestones Museum, Basingstoke on 6th September 2009.

A rare Farmers Foundry Portable No. 39, built 1910, on display at Holkham Hall on 15th December 2009. Only one other is known in the UK in the Saunders Collection.

Engine Cross-reference list

Notes: *This listing is purely to cross-reference engines that are in the main directory, only the minimal details of each engine are itemised here. Look under the type of engine you are seeking and from its works number, registration or build date you can see where the engine is located. Refer to the main sections for further information.*

ALBARET
946	1923	Chatham Steam Centre, the Historic Dockyard

ALLEN - JOHN
67BW	4613	Hollycombe House - Steam in the Country

ARMSTRONG - WHITWORTH
10R22	DX 4602	Lowestoft, East Anglian Transport Museum

AVELING - BARFORD
AD 073	CUP 396	Newby Bridge, Lakeside & Haverthwaite Railway
AG 758	FRM 973	Amberley Chalk Pits Museum
AH162	JXH 174	Chatham Steam Centre, the Historic Dockyard
AH 363	CTL 225	Beamish Open Air Museum
AH 412	ECT 452	Cardiff, Welsh Industrial & Maritime Museum, St. Fagans

AVELING & PORTER
		Sittingbourne & Kemsley Light Railway, Kemsley Station
721	KL 3631	South Kensington, Science Museum
803	Rail loco	Aylesbury, Quainton Road
2185	BY 2185	South Kensington, Science Museum
2992	AB 9331	Birmingham Science Museum, Dollman Street
3319	HR 6013	Leicester Museum of Technology
3567	Rail loco	Chatham Steam Centre, the Historic Dockyard
4735	PC 1414	Launceston Steam Railway
5156	TU 874	Chatham Steam Centre, the Historic Dockyard
6221	EB 02	Strumpshaw Steam Museum
6530	NM 291	Luton, Stockwood Museum
8097	FX 7014	Chatham Steam Centre, the Historic Dockyard
8169	KT 998	Fakenham, Thursford Collection
8178	PU 700	Fakenham, Thursford Collection
8200	H 0778	Fakenham, Thursford Collection
8370	NK 4562	Swansea Waterfront Museum
8563	IY 1682	Stradbally Steam museum. Laois, Eire
8653	KE 6455	Hollycombe House - Steam in the Country
8778	E 5359	Chesterfield, Barrow Hill Roundhouse Railway Centre
8890	DO 1943	Fakenham, Thursford Collection
8891	DO 1944	Fakenham, Thursford Collection
8906	DE 5880	Cardiff, Welsh Industrial & Maritime Museum, St. Fagans
9010	AP 9235	Fakenham, Thursford Collection
9036	BP 6065	Fakenham, Thursford Collection
9149	KE 2202	Fakenham, Thursford Collection
9449	Rail loco	Chinnor & Princes Risborough Railway
10003	AH 6130	Fakenham, Thursford Collection
10050	XM 4037	Hollycombe House, Steam in the Country
10079	BJ 7116	Strumpshaw Steam Museum

10143	FX 9683	Hollycombe House, Steam in the Country
10271	ME 2103	Chatham Steam Centre, the Historic Dockyard
10312	BK 674	Portsmouth City Museum
10317	FX 9412	Dorchester, Recreational Park
10324	NO 5898	Fakenham, Thursford Collection
10341	PW 623	Fakenham, Thursford Collection
10342	NO 5896	Fakenham, Thursford Collection
10345	NO 5893	Fakenham, Thursford Collection
10347	NO 5891	Fakenham, Thursford Collection
10399	BH 9624	Chatham Steam Centre, the Historic Dockyard
10415	ME 5770	Fakenham, Thursford Collection
10437	OS 1314	Fakenham, Thursford Collection
10456	NO 7239	Fakenham, Thursford Collection
10471	NM 2570	Luton, Stockwood Museum
10617I	K 7224	Dublin, The National Transport Museum, Howth Castle
10626	DL 3128	Brading, Isle of Wight, 'The Experience'
10675	PR 1070	Fakenham, Thursford Collection
10755	PD 1510	Fakenham, Thursford Collection
10780	PR 1392	Fakenham, Thursford Collection
10784	PU 3094	Sittingbourne & Kemsley Light Railway, Kemsley Station
10785	PD 7738	Fakenham, Thursford Collection
11055	NY 6931	Dover Transport Museum
11205	MO 5549	Fakenham, Thursford Collection
11322	PT 6445	Fakenham, Thursford Collection
11454	PE 9234	Fakenham, Thursford Collection
11648	RT 2474	Lowestoft, East Anglian Transport Museum
11822	UF 1934	Fakenham, Thursford Collection
11918	YT 4531	Fakenham, Thursford Collection
11980	DY 4878	Fakenham, Thursford Collection
12181	DW 6156	Fakenham, Thursford Collection
12186	PK 2684	Fakenham, Thursford Collection
12205	VF 4393	Fakenham, Thursford Collection
12461	MY 953	Strumpshaw Steam Museum
14008	VG 2269	Fakenham, Thursford Collection
14025	HF 6621	Liverpool, Museum of Liverpool
14044	DW 7088	Chatham Steam Centre, the Historic Dockyard
14072	DV 7077	Chatham Steam Centre, the Historic Dockyard
14073	FV 1326	Chatham Steam Centre, the Historic Dockyard
14085	VF 9891	Strumpshaw Steam Museum
14090	NG 564	Amberley Chalk Pits Museum
14121	FG 7099	Alfold Grampian Transport Museum
14135	NG 1412	Strumpshaw Steam Museum
14137	TM 9357	Fakenham, Thursford Collection
14163	MJ 4597	Fakenham, Thursford Collection

BABCOCK & WILCOX
95/4011	YB 7978	Bicton Gardens, Devon

BROUHOT
-	Portable	Hollycombe House, Steam in the Country

BROWN & MAY
6226	Portable	Chippenham, Lackham Agricultural & Rural Life Trust
6691	Portable	Hollycombe House, Steam in the Country

BURRELL

748	CF 3667	Thetford, Charles Burrell Museum
776	-	Stowmarket, Museum of East Anglian Life
777	-	Stowmarket, Museum of East Anglian Life
1876	AH 5366	Hollycombe House, Steam in the Country
2363	Portable	Diss, Bressingham Steam Museum,
2366	AH 5488	Strumpshaw Steam Museum
2479	AH 5239	Thetford, Charles Burrell Museum
2551	-	Sittingbourne, Bredgar & Wormshill Railway
2626	TB 3717	Thetford, Charles Burrell Museum
2706	TB 2845	Draycott-in-the-Clay, Klondyke Mill
2780	CL 4300	Fakenham, Thursford Collection
2877	HT 3163	Exmouth, Sandy Bay, World of Country Life,
2948	NO 106	Thetford, Charles Burrell Museum
3075	CL 4301	Fakenham, Thursford Collection
3098	MA 5657	Liverpool, Museum of Liverpool
3106	NO 439	Thetford, Charles Burrell Museum
3112	CF 3440	Diss, Bressingham Steam Museum
3200	CL 4296	Fakenham, Thursford Collection
3305	TB 3722	Fleetwood, Farmer Parrs Animal World
3356	Steam yachts	Hollycombe House, Steam in the Country
3399	AH 5446	Stowmarket, Museum of East Anglian Life
3444	CK 3403	Scarborough, The Scarborough Fair Collection
3509	G 6469	Launceston, Dingles Fairground Heritage Centre
3526	AO 6302	Northallerton, Prestons of Potto
3545	AH 0173	Hollycombe House, Steam in the Country
3618	TD 4276	Northallerton, Prestons of Potto
3622	NO 306	Thetford, Charles Burrell Museum
3695	TB 2740	Thetford, Charles Burrell Museum
3711	TA 3067	Exmouth, Sandy Bay, World of Country Life,
3786	PB 9610	Brighton, British Engineerium
3789	SW 742	Strumpshaw Steam Museum
3815	BP 5757	Hollycombe House, Steam in the Country
3827	CL 4299	Fakenham, Thursford Collection
383	3BR 1498	Thetford, Charles Burrell Museum
3850	AH 0775	Fakenham, Thursford Collection
3878	T 3997	Scarborough, The Scarborough Fair Collection
3884	H 5728	Exmouth, Sandy Bay, World of Country Life,
3924	O 6085	Strumpshaw Steam Museum
3962	PW 1714	Diss, Bressingham Steam Museum
3986	AF 9293	Launceston, Dingles Fairground Heritage Centre
3993	CF 5646	Diss, Bressingham Steam Museum,
3996	E 9597	Diss, Bressingham Steam Museum,
4032	PW 6287	Thirsk, Ampleforth Farming Flashback
4045	PW 8878	Fakenham, Thursford Collection
4053	TD 8047	Fordingbridge, Breamore House
4061	OU 945	Thetford, Charles Burrell Museum
4084	FK 3564	Birmingham Science Museum, Dollman Street

CHATHAM

-	Steam boiler	Chatham Steam Centre, the Historic Dockyard

CLARKE

-	Portable Barnsley,	Wortley Top Forge

CLAYTON & SHUTTLWORTH
13818	Portable	Beamish Open Air Museum
15635	Portable	Reading, Museum of English Rural Life
44140	Portable	Hollycombe House, Steam in the Country
48308	FE 2754	Fakenham, Thursford Collection
49105	TL 555	Fakenham, Thursford Collection
50010	Portable	Hollycombe House, Steam in the Country

CROSKILL
-	Portable	Nottingham, Woollaton Park

DAVEY PAXMAN
-	Portable	Southampton, Bursledon Brickworks
UP9	Portable	Hodgehill Nurseries, Kidderminster

FARMERS FOUNDRY
39	Portable	Holkham Hall Bygones Museum, Well-next-the-Sea

FODEN
848	DD 4894	Birmingham Science Museum, Thinktank
1174	SN 1609	Wroughton, Science Museum
6368	M 8118	Wroughton, Science Museum
10788	AV 6695	Leyland, British Commercial Vehicle Museum
11892	DD 7916	Kendal, Levens Hall
13358	DF 8187	Fakenham, Thursford Collection
13708	VF 8862	Diss, Bressingham Steam Museum
13802	VN 2912	Strumpshaw Steam Museum

FOSTER
2821	BE 7448	Diss, Bressingham Steam Museum
3643	Light m/c	Northallerton, Prestons of Potto
14422	CT 6708	Weybourne, North Norfolk Railway
14438	NR 7262	Belfast, Ulster Folk & Transport Museum

FOWLER
6188	MA 8528	Diss, Bressingham Steam Museum
7232	WR 6662	Sheffield, Kelham Island Museum
7769	HC 2431	Brading, Isle of Wight, 'The Experience'
9971	HO 5655	Exmouth, Sandy Bay, World of Country Life,
10318	WR 6790	Leyland, British Commercial Vehicle Museum
13482	NO 797	Tamworth, Statfold Barn Railway, Staffordshire
14383	NM 1284	Hollycombe House, Steam in the Country
14728	DO 1918	Barleylands Farm Museum
14729	DO 1919	Barleylands Farm Museum
15116	CB 3062	Kendal, Levens Hall
15194	FX 6820	Wroughton, Science Museum
15195	FX 6821	Wroughton, Science Museum
15337	DO 1928	Draycott-in-the-Clay, Klondyke Mill
15340	BL 8167	Strumpshaw Steam Museum
15341	BL 8168	Strumpshaw Steam Museum
15428	DD 1994	Leeds, Armley Mills Industrial Museum
15429	DD 1995	Leeds, Armley Mills Industrial Museum
15490	CA 6328	Beamish Open Air Museum
15657	FX 6661	Scarborough, The Scarborough Fair Collection

15662	AZ 2450	Belfast, Ulster Folk & Transport Museum
15698	BW 6179	Chatham Steam Centre, the Historic Dockyard
15732	PT 832	Chatham Steam Centre, the Historic Dockyard
15787	EP 2398	Draycott-in-the-Clay, Klondyke Mill
16053	EB 5518	Lincoln, Museum of Lincolnshire Life
16054	EB 5519	Lincoln, Museum of Lincolnshire Life
16402	SW 1854	Aberfeldy Park
17251	DS 7206	Coatbridge, Summerlee, Scottish Museum of Industrial Life
17621	-	Leeds, Armley Mills Industrial Museum
17756	VO 8988	Nottingham, Woollaton Park
17757	VO 8987	Nottingham, Woollaton Park
18507	SM 8832	Draycott-in-the-Clay, Klondyke Mill
18877	WX 6358	Beamish Open Air Museum
19357	CZ 2260	Belfast, Ulster Folk & Transport Museum
19590	JI 5869	Belfast, Ulster Folk & Transport Museum
21649	KI 1973	Stradbally Steam Museum, Laois, Eire

The Long Shop Museum in Leiston, Suffolk was the old Garrett works and currently displays a range of Garrett engines as well as many other items of relevant machinery and small objects. One resident is Garrett 44180 'The Joker' the only survivor of four direct plough engines built to this design. Seen here at Knowle Hill Rally on 12th August 1973.

GARRETT

-	Portable	Dereham, Gressenhall Farm & Workhouse Museum
29764	AC 9326	Southampton, Bursledon Brickworks
30877	Portable	Leiston, Long Shop Museum
32944	BJ 3206	Leiston, Long Shop Museum
33180	BJ 4483	Leiston, Long Shop Museum

33284	DX 3099	Scarborough, The Scarborough Fair Collection
33305	HT 7112	Sittingbourne, Bredgar & Wormshill Railway
33348	SZ 100	Hollycombe House, Steam in the Country
33442	BL 9009	Sittingbourne, Bredgar & Wormshill Railway
33818	BJ 5323	Brading, Isle of Wight, 'The Experience'
33902	BJ 5340	Fakenham, Thursford Collection
34045	BJ 6628	Strumpshaw Steam Museum
34187	AH 9623	Fakenham, Thursford Collection
34193	Portable	Launceston, Dingles Fairground Heritage Centre
34265	SV 4945	Leiston, Long Shop Museum
34471	Portable	Leiston, Long Shop Museum
34641	CF 5913	Diss, Bressingham Steam Museum

GREEN, T & SONS

1978	U 9647	Bury, East Lancashire Railway Museum
2054	RF 3309	Draycott-in-the-Clay, Klondyke Mill

GRENVILLE

Steam carriage	875	Beaulieu, National Motor Museum

HOUSTON

-1	985	Coatbridge, Summerlee, Scottish Museum of Ind. Life

LAWSON

Lawson	SA 16	Alfold Grampian Transport Museum

LEYLAND

-	1923	Leyland, British Commercial Vehicle Museum

MANN

1216	U 4249	Stradbally Steam Museum, Laois, Eire
1260	U 4298	Hollycombe House, Steam in the Country
1747	MUP 662	Beamish Open Air Museum

MARSHALL

-	Portable	Brighton, British Engineerium
UP 26	Portable	Blaenafon, Big Pit National Coal Museum
-	Portable	Brentwood, Old MacDonald's Farm
-	Portable	Milton Keynes Rural Life Museum
-	Portable	Angmering, Haskins Garden Centre
20146	Portable	Chatham Steam Centre, the Historic Dockyard
23368	TA 2758	Bicton Gardens, Devon
26660	Portable	Caister Castle Museum
28334	EB 6519	Strumpshaw Steam Museum
30169	Portable	Lincoln, Museum of Lincolnshire Life
31910	Portable	Dublin, Straffan Steam Museum
46699	Portable	Draycott-in-the-Clay, Klondyke Mill
47731	MS 3081	Edinburgh, National Museums Collections Centre, Granton
52483	E 5114	Strumpshaw Steam Museum
53755	D 5453	Bewick-on-Tweed, Chain Bridge Honey Farm
59393	Portable	Draycott-in-the-Clay, Klondyke Mill
66182	CT 3926	Strumpshaw Steam Museum
68823	BE 3044	Coalville, Snibston Discovery Park

68872	OT 3092		Amberley Chalk Pits Museum
69872	Portable		Arreton Barns Craft Village & Museum
71396	Portable		Bury, East Lancashire Railway Museum
78539	J 17884		Jersey, Pallot Steam, Motor & General Museum
79108	PW 5042		Fakenham, Thursford Collection
79669	PX 2690		Amberley Chalk Pits Museum
80608	FG 1191		Amberley Chalk Pits Museum
81427	PY 6079		Draycott-in-the-Clay, Klondyke Mill
81451	-		Jersey, Pallot Steam, Motor & General Museum
83270	VH 1261		Strumpshaw Steam Museum
83600	VF 3665		Strumpshaw Steam Museum
83780	VF 4183		Strumpshaw Steam Museum
84480	Portable		Southampton, Bursledon Brickworks
84679	WF 2657		Tamworth, Statfold Barn Railway, Staffordshire
85601	SC7488		Dunfermline, Lathalmond Bus Museum
86753	-		Bicton Gardens, Devon
87087	PO 7927		Brading, Isle of Wight, 'The Experience'
88553	Portable		Westonzoyland Pumping Station
89200			Alfold Grampian Transport Museum

MARSHALL - J.T.

-	1886		PortableNottingham, Woollaton Park

McLAREN

112	KL2176		Wroughton, Science Museum

MERLIN & CIE

-	1924	Portable	Jersey, Pallot Steam, Motor & General Museum
-	?	In build	Jersey, Pallot Steam, Motor & General Museum
-	-		PortableSt Austell, alongside the A390
-	-	Portable	Swaffham, Cockley Cley, Iceni Village

MERRYWEATHER

Valiant type	-	FP	Banbury, Bygones Museum
Trench steamer	-		Sheffield Fire & Police Museum
-	-	FE	Belfast, Cultra, Ulster Folk & Trspt Museum
Sutherland	1863	FE	Wroughton, Science Museum
Middleton Distillery	1880	FE	Midleton, The Old Distillery, Co. Cork, Ireland
Gateshead	1880?	FE	York, National Railway Museum
-	1885	FE	Exmouth, Sandy Bay, World of Country Life
Bexhill	1885	FE	Weedon, National Fire Service Museum
Prudhoe UDC 3254	1887	FE	Newcastle Discovery Museum
GNR Dundalk Wks	1889	FE	Dublin, The National Transport Museum, Howth Castle
Penrhyn Castle	17881894	FE	Bangor, Penrhyn Castle Museum
Victoria/Oxford	1900?	FE	Oxford City Fire & Rescue Service
Queen Victoria	42961901	FE	Leiston, Long Shop Museum
1952 Valiant	1901	FP	York, National Railway Museum
John Brown.	18181901	FE	Glasgow Museum of Transport
Bridport Fire Brig.	1902	FE	Bridport, Highlands End Holiday Park
-	1902	FE	Maidstone Museum & Bentlif Art Gallery
LCC Fire Brigade	1902	FE	Wroughton, Science Museum
Newport Pagnell	1912	FE	Milton Keynes Rural Life Museum
Tenbury	1912	FE	Swansea, Welsh Museum of Fire
1292	-	FE	Basingstoke, Milestones Museum

-	1914	FP	Diss, Bressingham Steam Museum
-	Valiant 1915?	FP	Beaulieu, National Motor Museum
2142 Valiant	-	FP	Saltash, St. Dominick, Cotehele House,
2428	-	FP	Doncaster, Cusworth Hall and Park
2662	-	FE	Swansea, South Wales Fire & Rescue Service
Colstoun Hse 2743	1908	FE	Hollycombe House, Steam in the Country
3702	1914	FE	Diss, Bressingham Steam Museum
4716	-	FP	Brownhills, Chasewater Railway (At Brownhills West Station)
8558. B785	-	FP	Brighton, British Engineerium
The York St. Flax	-	FE	Holywood, County Down, Ulster Folk & Trspt Museum
-	1940?	FP	Peastonbank, Glenkinchie distillery

RANSOMES, SIMS & JEFFERIES

E409	Portable	Brighton, British Engineerium
5137	Z1	Stowmarket, Museum of East Anglian Life
14329	Portable	Draycott-in-the-Clay, Klondyke Mill
15740	J 14457	Jersey, Pallot Steam, Motor & General Museum
31136	DM 3048	Cardiff, Welsh Industrial & Maritime Museum, St. Fagans
36030	Portable	Strumpshaw Steam Museum
42019	Portable	Barleylands Farm Museum

ROBEY

-	Portable	Tavistock, The Robey Trust
-	Steam producer	Harlington, Beds, Poplars Garden Centre
31821	Tr	Tavistock, The Robey Trust
32387	DE 2592	Tavistock, The Robey Trust
33348	Z 200	Tavistock, The Robey Trust
33810	Portable	Hollycombe House, Steam in the Country
40062	Portable	Fordingbridge, Breamore House
40991	FE 5373	Tavistock, The Robey Trust
42520	FE 7632	Diss, Bressingham Steam Museum
42693	FE 7490	Tavistock, The Robey Trust
43165	FE 9350	Exmouth, Sandy Bay, World of Country Life,
45655	VL 2773	Tavistock, The Robey Trust
51407	Exploded boiler	Beamish Open Air Museum
53445	Portable	Fochabers, Christies Garden Centre
57193	Steam Producer	Tavistock, The Robey Trust

RICHARD HORNSBY

-	Portable	Wroughton, Science Museum
2598	Portable	Beamish Open Air Museum
7297	TE	Skegness, Church Farm Museum

RUSTON PROCTOR

10743	Portable	Lincoln, Museum of Lincolnshire Life
18188	Portable	Birmingham Science Museum, Thinktank
30656	Portable	Hollycombe House, Steam in the Country
38994	AH 7938	Strumpshaw Steam Museum
39872	RT 1487	Fakenham, Thursford Collection
46596	SFE 519L	Lincoln, Museum of Lincolnshire Life
47319	-	Beamish Open Air Museum
51457	-	Blackgang Chine, Nr, Ventnor, Isle of Wight

RUSTON & HORNSBY

-	Portable	Sheffield, Renishaw Hall
113812	ME 3273	Glasgow Museum of Transport
115023	XM 6373	Sittingbourne, Bredgar & Wormshill Railway
122282	BJ 9644	Lowestoft, East Anglian Transport Museum
169166	TL 3433	Northallerton, Prestons of Potto

SAVAGE

No.3	OE	Denby Dale, Sisset, Nortonthorpe Mills
418	OE	Fakenham, Thursford Collection
421	CE	Fakenham, Thursford Collection
449	OE	Fakenham, Thursford Collection
561	CE	Ramsey Rural Museum, Cambs
607	CE	Scarborough, The Scarborough Fair Collection
627	CE	Beamish Open Air Museum
713	OE	Beamish Open Air Museum
739	OE	Birmingham Science Museum
740	OE	Fakenham, Thursford Collection
761	Light m/c	Northallerton, Prestons of Potto
762	CE	Fakenham, Thursford Collection
763	OE	Fakenham, Thursford Collection
772	CE	Hollycombe House, Steam in the Country
869	CE	Hollycombe House, Steam in the Country

SENTINEL

753	V 3057	Alfold Grampian Transport Museum
1286	AW 2964	Glasgow Museum of Transport
1488	AW 3407	Burton, Coors Museum
1718	AW 3835	Wroughton, Science Museum
5676	EC 5927	Coatbridge, Summerlee, Museum of Industrial Life
6842	KA 6147	Liverpool, Museum of Liverpool
9208	BYL 485	Chatham Steam Centre, the Historic Dockyard

SHAND MASON

P3	-FE	Southwark, London Fire Brigade Museum
1127	1875 FE	Morton-in-the-Marsh Fire Service College
GNSR Fire Brigade	-FE	Alfold Grampian Transport Museum
-	1876 FE	Stoke Potteries Museum & Art Gallery
Fulwood UDC	1880 FE	Preston, Lancashire Fire & Rescue Service
City of Chester	1880? FE	Chester Fire & Rescue Service
Carrow Works	1881 FE	Norwich, Bridewell Museum
Horrock's Norwich	1882 FE	Sheffield Fire & Police Museum
Bexhill	1885 FE	Weedon, National Fire Service Museum
Shepshed	1889 FE	Coalville, Snibston Discovery Park
-	1890 FP	Kenilworth, Stoneleigh Abbey
The Nelson	1890 FE	Beamish Open Air Museum
Tatton	1890 FE	Knutsford, Tatton Hall
Victoria No.48	1891 FP	Southwark, London Fire Brigade Museum
Rickmansworth	1891 FE	Southwark, London Fire Brigade Museum
City of Leeds	1891 FE	Leeds, Armley Mills Industrial Museum
Victor	1892 FE	Poole Museum
Bor' of Barnstable	1892 FE	Brighton, British Engineerium
Englefield	1894 FE	Reading, Englefield Estate & Museum
Southgate	1894 FE	Wroughton, Science Museum

J.J. Coleman	1895 FP	Diss, Bressingham Steam Museum
Hertfordshire F.B	1896 FE	Watford Fire Station & Museum
Hazelmere	1896 FE	Enfield, Whitewebbs Museum of Transport
Warnham Court	1896 FE	Amberley Chalk Pits Museum
Arundel Castle	1895 FE	Amberley Chalk Pits Museum
Glenlossie	1895? FE	Forres, Dallas Dhu Historic Distillery
-	Steam Pump	Barnsley, Wortley Top Forge
-	1897 FE	Maidstone, Kent Fire Service
-	1898 FE	Birmingham Science Museum, Dollman Street
Normanby House	1900 FE	Scunthorpe, Normanby Hall Country Park
Florian	15921901 FE	Birmingham, West Midlands Fire & Rescue Service
N. Lancs Fire Force	1901 FE	Liverpool, Museum of Liverpool
Dunrobin Castle	1903 FE	Dunrobin Castle
Bor' of Huntingdon	1903 FE	Morton-in-the-Marsh Fire Service College
Arkle	1904 FE	York Castle Museum
Hinton Admiral	1904 FE	Fordingbridge, Breamore House
BSA ll	1906 FE	Birmingham Science Museum, Dollman Street
City of York	1907 FE	Beamish Open Air Museum
-	Incomplete FP	Birmingham Science Museum, Dollman Street
Carrickfergus	1908 FE	Carrickfergus Museum & Civic Centre
Thorney 2015	1908 FE	Biggleswade Fire Station, Bedfordshire
Francis Dudley	1909 FE	Kenilworth, Stoneleigh Abbey
George V	1910 FE	Rochdale, Greater Manchester Fire Service Museum
Needham Market	1911 FE	Ipswich, Suffolk Fire Service
Ely Fire Station	1912 FE	Ely Fire Station
Cliveden	- FE	Helston, Flambards, the Experience
Lymington	- FE	Winchester, Chilcomb House

TASKER

111	Semi portable	Basingstoke, Milestones Museum
203D	Stationary Boiler	Winchester, Chilcomb House
352	NO 1060	Winchester, Chilcomb House
1228	Portable	Basingstoke, Milestones Museum
1235	Portable	Calbourne Mill, Isle of Wight
1352	HO 5600	Basingstoke, Milestones Museum
1396	AA 2254	Amberley Chalk Pits Museum
1599	AA 5296	Basingstoke, Milestones Museum
1643	AA 5506	Winchester, Chilcomb House, (parts only)
1643	AA 5506	Basingstoke, Milestones Museum. (parts only)
1675	AP 9027	Winchester, Chilcomb House
1715	BP 6289	Winchester, Chilcomb House
1726	SR 1294	Basingstoke, Milestones Museum
1776	AP 9281	Winchester, Chilcomb House
1818	HO 2822	Basingstoke, Milestones Museum
1906	BD 7994	Basingstoke, Milestones Museum
1915	YB 183	Basingstoke, Milestones Museum
1933	OT 8201	Basingstoke, Milestones Museum

THORNYCROFT

No.1	1896	Leyland, British Commercial Vehicle Museum
115	EL 3908	Basingstoke, Milestones Museum

TIDMAN
1870?	CE	Holkham Hall Bygones Museum, Well-next-the-Sea
1881	CE	Hollycombe House, Steam in the Country
1881	OE	Hollycombe House, Steam in the Country
1885	Portable	Dereham, Gressenhall Farm & Workhouse Museum
1890	OE	Norwich, Bridewell Museum
1891	OE – Carousel	Diss, Bressingham Steam Museum
1891	CE – Carouse	lDiss, Bressingham Steam Museum
1900	OE	Wroughton, Science Museum
1900	CE	Wroughton, Science Museum

TROTTER
-	1933 Vb RR	Gloucester City Museum

TURNER
981	Portable	Stowmarket, Museum of East Anglian Life

TUXFORD
1131	Portable	Lincoln, Museum of Lincolnshire Life
1234	Portable	Edinburgh, National Museums Collections Centre, Granton

WALKER
1910	CE	Hollycombe House, Steam in the Country

WALLIS & STEEVENS
2660	CJ 4816	Ironbridge Gorge Museum, Blists Hill
2931	AA 2167	Liverpool, Museum of Liverpool
7279	AA 2470	Basingstoke, Milestones Museum
7614	AF 3456	Winchester, Chilcomb House
7650	DL 2864	Brading, Isle of Wight, 'The Experience'
7801	PU 3378	Strumpshaw Steam Museum
7867	OT 3078	Basingstoke, Milestones Museum
7939	OT 6514	Tamworth, Statfold Barn Railway, Staffordshire
7940	OT 8512	Basingstoke, Milestones Museum
7986	OT 8207	Bradford Industrial Museum
8023	CG 571	Hollycombe House, Steam in the Country
8033	OU 5185	Chatham Steam Centre, the Historic Dockyard
8104	GOR 248	Amberley Chalk Pits Museum

WALSH & CLARK
5038	-	Stowmarket, Museum of East Anglian Life
5039	-	Stowmarket, Museum of East Anglian Life
-	-	Wroughton, Science Museum

WATERLOO
1666	-	Exmouth, Sandy Bay, World of Country Life,

WILLIAM ROSE
1892	FE	Belfast, Ulster Folk & Transport Museum
Georgina Rose	FE	Burton-on-Trent, Staffordshire Fire & Rescue Museum

YOUNGS
Youngs	Portable	Diss, Bressingham Steam Museum

Museums by County

ENGLAND

AVON
Bristol Industrial Museum

BEDFORDSHIRE
Biggleswade Fire Station
Harlington, Poplars Garden Centre
Luton, Stockwood Museum

BERKSHIRE
Reading, Englefield Estate
Reading, Museum of English Rural Life.

BUCKINGHAMSHIRE
Aylesbury, Quainton Road
Milton Keynes Rural Life Museum
Ely Fire Station,
Ramsey Rural Museum

CAMBRIDGESHIRE
Cambridge Museum of Technology

CHESHIRE.
Chester, Cheshire Fire & Rescue Service
Knutsford, Tatton Hall

CORNWALL
Helston, Flambards, the Experience
Launceston, Dingles Fairground Heritage Centre
Launceston Steam Railway
St Austell, on private land alongside the A390
Saltash, St. Dominick, Cotehele House

CUMBRIA
Kendal, Levens Hall,
Newby Bridge, Lakeside & Haverthwaite Railway

DERBYSHIRE
Chesterfield, Barrow Hill Roundhouse Railway Centre

DEVON
Bicton Gardens
Exmouth, Sandy Bay, World of Country Life
Tavistock, The Robey Trust

DORSET
Bridport, Highlands End Holiday Park
Dorchester, Recreational Park
Poole Museum, Poole

DURHAM
Beamish Open Air Museum

ESSEX
Barleylands Farm Museum
Brentwood, Old MacDonald's Farm

GLOUCESTERSHIRE
Gloucester City Museum

GREATER MANCHESTER
Rochdale, Greater Manchester Fire Service Museum

HAMPSHIRE
Basingstoke, Milestones Museum
Beaulieu, National Motor Museum
Fordingbridge, Breamore House
Portsmouth City Museum
Southampton, Bursledon Brickworks
Winchester, Chilcomb House

HERTFORDSHIRE
Watford Fire Station & Museum

ISLE OF WIGHT
Arreton Barns, Craft Village & Museum.
Blackgang Chine, Nr. Ventnor
Brading, Isle of Wight, 'The Experience'
Calbourne Mill

KENT
Canterbury, Preston Services
Chatham Steam Centre, the Historic Dockyard, Medway
Dover Transport Museum
Maidstone, Kent Fire Service
Maidstone Museum & Bentlif Art Gallery
Sittingbourne & Kemsley Light Railway, Kemsley Station
Sittingbourne, Bredgar and Wormshill Railway

LANCASHIRE
Bury, East Lancashire Railway Museum
Fleetwood, Farmer Parrs Animal World
Leyland, British Commercial Vehicle Museum,
Preston, Lancashire Fire & Rescue Service

LINCOLNSHIRE
Lincoln, Museum of Lincolnshire Life
Scunthorpe, Normanby Hall Country Park
Skegness, Church Farm Museum

LEICESTERSHIRE
Coalville, Snibston Discovery Park
Leicester Museum of Technology

LONDON
Enfield, Whitewebbs Museum of Transport
Kew Bridge Steam Museum
Southwark, London Fire Brigade Museum
South Kensington, Science Museum

MERSEYSIDE
Liverpool, Museum of Liverpool

Traction engines have been around for a long time, many, work-weary, were retired and lay derelict for years. Here at Cushing's of Fakenham in Norfolk three quietly rust away being overwhelmed by encroaching vegetation. Burrell Showman's No.2780, 'King Edward VII' with an Aveling & Porter behind and yet a further engine almost hidden by the trees in the background. Amazingly engines in this condition are often resurrected 'phoenix' style to working order or to be lovingly cared for in museums. This photo was taken in August 1972.

NORFOLK
Caister Castle Museum
Dereham, Gressenhall Farm & Workhouse Museum
Diss, Bressingham Steam Museum
Fakenham, Thursford Collection
Norwich, Bridewell Museum
Strumpshaw Steam Museum
Swaffham, Cockley Cley, Iceni Village
Thetford, Charles Burrell Museum
Wells-next-the-Sea, Holkham Hall Bygones Museum
Weybourne, North Norfolk Railway
NORTHAMPTONSHIRE
Weedon, National Fire Service Museum

NORTHUMBERLAND
Berwick-on-Tweed, Chain Bridge Honey Farm

NOTTINGHAMSHIRE
Nottingham, Woollaton Park

OXFORDSHIRE
Banbury, Bygones Museum
Chinnor & Princes Risborough Railway
Didcot Railway Centre
Moreton-in-the-Marsh, Fire Service College
Oxford City Fire & Rescue Service

SHROPSHIRE
Ironbridge Gorge Museum, Blists Hill

SOMERSET
Westonzoyland Pumping Station

STAFFORDSHIRE
Burton-on-Trent, Coors Museum
Burton-on-Trent, Staffordshire Fire & Rescue Service Museum
Draycott-in-the-Clay, Klondyke Mill
Stoke, Potteries Museum & Art Gallery
Tamworth, Statfold Barn Railway

SUFFOLK
Ipswich, Suffolk Fire Service
Leiston, Long Shop Museum.
Lowestoft, East Anglian Transport Museum
Stowmarket, Museum of East Anglian Life

SUSSEX
Amberley Chalk Pits Museum
Angmering, Haskins Garden Centre
Brighton, British Engineerium
Hollycombe House, Steam in the Country

TYNE & WEAR
Newcastle Discovery Museum

WARWICKSHIRE
Kenilworth, Stoneleigh Abbey

WEST MIDLANDS
Birmingham Science Museum, Dolman Street
Birmingham Science Museum, Thinktank
Birmingham, West Midlands Fire & Rescue Service
Brownhills, Chasewater Railway

WILTSHIRE
Chippenham, Lackham Agricultural & Rural Life Trust
Wroughton, Science Museum, Swindon,

WORCESTERSHIRE
Kidderminster, Hodgehill Nurseries

YORKSHIRE
Barnsley, Wortley Top Forge
Bradford Industrial Museum
Denby Dale, Sisset, Nortonthorpe Mills
Doncaster, Cusworth Hall and Park
Leeds, Armley Mills Industrial Museum
Northallerton, Prestons of Potto, Scarborough,
Scarborough, The Scarborough Fair Collection
Sheffield Fire & Police Museum
Sheffield, Kelham Island Museum
Sheffield, Renishaw Hall
Thirsk, Ampleforth Farming Flashback
York Castle Museum
York, National Railway Museum

SCOTLAND

ABERDEENSHIRE
Alfold, Grampian Transport Museum

EAST LOTHIAN
Peastonbank, Glenkinchie Distillery
FIFE
Dunfermline, Lathalmond Bus Museum

GLASGOW
Glasgow Museum of Transport

HIGHLAND
Dunrobin Castle, Golspie

LANARKSHIRE
Coatbridge, Summerlee, Scottish Museum of Industrial Life

MIDLOTHIAN
Edinburgh, Lothian and Borders Fire and Rescue Service
Edinburgh, National Museums Collections Centre, Granton
Edinburgh, Royal Museum of Scotland

MORAYSHIRE
Fochabers, Christies Garden Centre
Forres, Dallas Dhu Historic Distillery

PERTH & KINROSS
Aberfeldy, Victoria Park

STRATHCLYDE
Greenock, Strathclyde Fire Preservation Group

WALES

CARDIFF
Cardiff, Welsh Industrial & Maritime Museum, St. Fagans

GWYNEDD
Bangor, Penrhyn Castle Museum

SWANSEA
Swansea, South Wales Fire & Rescue Service
Swansea Waterfront Museum
Swansea, Welsh Museum of Fire

TORFAEN
Blaenafon, Big Pit National Coal Museum

NORTHERN IRELAND

Belfast, Ulster Folk & Transport Museum, Co. Antrim
Carrickfergus Museum & Civic Centre, Co. Antrim

EIRE

Dublin, Straffan Steam Museum, Co. Dublin
Dublin, The National Transport Museum, Howth Castle, Co. Dublin
Midleton Old Distillery, Co. Cork
Stradbally Steam Museum, Laois

CHANNEL ISLANDS

Jersey, Pallot Steam, Motor & General Museum

Marshall Roller, Reg: PY 6079 built in 1926 as works No. 81427 and named 'Anne' is seen here at the Midland Festival of Steam on 15th July 1972. She can normally be found at the Klondyke Mill at Draycott-in-the-Clay.

Further Orders

If you wish to obtain further copies of this directory please contact the author:
Barrie C. Woods, 124 Eastern Way, Letchworth Garden City, Herts. SG6 4PF.
Tel: 01462 - 634835, mobile: 07967 - 251090.
E.mail: bcwoods124@ntlworld.com.

Copyright: Barrie C. Woods, 12th December 2009.
For further information on this matter please contact:
**Barrie C. Woods, 124 Eastern Way, Letchworth Garden City. SG6 4PF.
01462-634835. e.mail:**

NOTES

This shot was taken on 22nd June 2008 of a remarkable little portable which was built by Tidman of Norwich in the 1880s; it weighs barely 1 ton and can be seen at the Gressenhall Farm & Workhouse Museum near Dereham in Norfolk.

About the author.

I have had a life-long fascination of steam from when I first sat at the top of the white cliffs at Hitchin on the East coast Main line, train-spotting with my pencil and notepad way back in 1959 at the tender age of just 11. Since those impressionable days and the inevitable demise of steam on the railways my interests spread to road vehicles and I first began visiting rallies, as well as steering and driving traction engines in the early 1970s. My mentor was a veteran of 86 years old by the name of 'Punch' Lawrence. This continued for some years, the engines we drove were the first two owned by the legendary Reg Saunders of Stotfold, near Hitchin. Today the sons, Ted & John carry on the Saunders tradition and have increased their steam fleet out of all proportion to those early days.

Work and family inevitably caused adjustments to be made to my activities for some years but never did the yen to get near steam diminish, indeed it continued to dominate my life as it does to this day. From 1973 my involvement in the Battle of Britain Locomotive Society, ensured a life-time of restoration work and fund-raising to get Bulleid Pacific 34081, named '92 SQUADRON' back into steam.

Later and now divorced, I set about travelling the world to capture the last remnants of 'real steam' rather than the UK preserved and main line activities which although very laudable, were not the same. Countries such as China, Cuba, Ukraine, South Africa, Zimbabwe and Colorado were all visited as well as nearer to home locations such as Poland and Germany, the former to drive engines on the main line with fare paying passengers - if only they knew who's hand was on the regulator!

Traction engine rallies have been high on the agenda for as long as I can remember and in the last couple of years a fascination with museum exhibits has come to the fore, hence this book. As to the future who knows? There is plenty of steam still around it just takes a little more patience these days to find! But while it's out there I'll be out there with my camera photographing it. **BCW.**